PROPERTY DEVELOPMENT

FOURTH EDITION

DAVID CADMAN AND ROSALYN TOPPING

E & FN SPON

An Imprint of Routledge

London and New York

First published 1978 by E & FN Spon, an imprint of Chapman & Hall
Second edition 1983
Reprinted 1989 (twice)
Third edition 1991
Reprinted 1991, 1993, 1994
Fourth edition 1995
Reprinted 1996, 1997

Reprinted 1998 by E & FN Spon, an imprint of Routledge
11 New Fetter Lane, London EC4P 4EE
29 West 35th Street, New York, NY 10001

Typeset in 10½/12½pt Garamond by Saxon Graphics Ltd
Printed and bound in Great Britain by St Edmundsbury Press Ltd,
Bury St Edmunds, Suffolk

British Library Cataloguing in Publication Data
A catalogue record for this book is available from the British Library

Library of Congress Cataloguing in Publication Data
A catalogue record for this book is available from the Library of Congress

ISBN 0–419–20240–4

∞ Printed on acid-free text paper, manufactured in accordance with ANSI/NISO
Z39.48–1992 and ANSI/NISO Z39.48–1984 (Permanence of Paper)

CONTENTS

SELECTED DEVELOPMENT SCHEMES

PREFACE

The First Edition of *Property Development* was published in 1977 and reflected upon the collapse of the property market in 1973/4. It opened with the following statement:

> The collapse of the property market . . . was a shattering experience for all those involved with property development and finance.

Three editions later, we could start with the same sentiment, a sobering thought for authors who in their first edition said:

> . . . we are anxious to contribute to a better understanding of the risks and pitfalls that face the developer . . .

Many of the problems and changes that we forecast have come to be: the institutional reassessment of the role of property within the portfolio; the growing acceptance of a more rapid rate of obsolescence for offices; the emergence and growth of out-of-town space; and, indeed, the changes in IT and telecommunications which were identified in the First Edition. Not least amongst these warnings was the forecast of a substantial overhang of space and lack of rental growth to which we drew attention in the Postscript of the last edition, a factor that, perhaps more than any other, affects market sentiment in early 1995. The real uncertainty and the need for a more measured approach therefore remains.

With this in mind, in this Fourth Edition, the new editor, Ros Topping, has taken the work forward, in particular with a much expanded chapter on the rationale of institutional investment, more detail on financial appraisal and an update on the planning system. In addition, with the help of my colleague Jayne Cox, the development process has been placed into an economic context and a new chapter has been provided on market research.

This book has always been intended as an introduction for students and others that come to the property development process for the first time. This remains its purpose and we have been greatly helped by the comments of those of you who work with students, especially at undergraduate level. One of your comments, in particular, has shaped the book. In this edition, the editor has included a number of case studies. We hope that these will help bring the text to life, and be a help to those of you that teach. We are grateful to the companies that gave us permission to use their projects and experiences.

David Cadman
BSc MA (Cantab) FRICS. CEDR Accredited Mediator
July 1995

ACKNOWLEDGEMENTS

Rosalyn would particularly like to extend an enormous thank you to Jayne Cox for writing both Chapter 7 on Market Research and the section on the Economic Context within Chapter 1, together with all her invaluable help and support.

Rosalyn and Jayne would also like to thank the following for their help with research and with the provision of material for the case studies: James Folkes, Richard Ludlow, Paul MacNamara, Guy Morrell, Adrian Lee, Dr Karen Siercaki, Andrew Smith, Brian Kemp, Mark Roberts, Simon Mullaly, David Phillips, Andrew Whitehorn, Mark Pennington, Will Cousins, Stephen Brown, Paul Clark, Martin Kemp and the developers who have contributed photographs to illustrate developments of the 1990s (Slough Estates plc, Burwood House Developments Ltd, Development Securities plc and Lynton plc).

Finally Rosalyn would like to thank her husband, Simon Chadwick, for his considerable input on Chapter 5, and both the Topping and Chadwick families for all their help with looking after Amy.

1 ———————————

INTRODUCTION: THE DEVELOPMENT PROCESS AND ITS ECONOMIC CONTEXT

1.1 INTRODUCTION

This book is an introduction to the process of property development in the UK. It is a practical book which describes the process enabling the interested reader to obtain a complete overview. Case studies are included to provide real illustrations of particular aspects of the development process. It is not intended to be a book for those who have already learnt the lessons of development by hard experience. It is intended to provide a basic understanding of a complex activity, from which further study can follow.

The term 'property development' can evoke many feelings depending on any particular viewpoint: it must be defined for the purposes of this book. Property development is a process that involves changing or intensifying the use of land to produce buildings for occupation. It is not the buying and selling of land for a profit: land is only one of the raw materials used. Others include building materials, infrastructure, labour, finance and professional services. For the purpose of this book, we are concerned with development that involves building activity for commercial use, whether carried out by the private or public sector.

Property development is an exciting, at times frustrating, complex activity involving the use of scarce resources. It is a high risk activity which often involves large sums of money tied up in the production process, providing a product which is relatively indivisible. The performance of the economy, at both national and local levels, directly influences the process. As the development process is often lengthy the assumptions made at the outset may have changed dramatically by completion. Success very often depends upon attention to the detail of the process and the quality of the judgement that guides it. Success, however, cannot be judged purely by the size of the profit or loss in financial terms. Property development is more than bricks and mortar: it brings together people each pursuing their own objectives and interests. As a result there are winners and losers whether measured in financial, aesthetic, emotional, social or other terms. For some, it can represent an unwelcome change, replacing the familiar with insensitive buildings and uses.

1.2 THE PROCESS

There are a variety of views on, and descriptions of, the development process (see Bibliography for further reading). At its most simple, property development can be likened to any other industrial production process that involves the combination of various inputs in order to achieve an output or product. In the case of property development, the product is a change of land use and/or a new or altered building in a process which combines land, labour, materials and finance. However, in practice the process is complex, often taking place over a considerable time period. The end product is unique, either in terms of its physical characteristics and/or its location. No other process operates under such constant public attention.

   The development process may be divided into the following main stages:
   1. Initiation
   2. Evaluation
   3. Acquisition
   4. Design and costing
   5. Permissions
   6. Commitment
   7. Implementation
   8. Let/manage/dispose

These stages may not always follow this sequence and often overlap or repeat. The sequence is typical of a speculative development where an occupier is not sought until the buildings have been completed. If the development is pre-sold to an occupier, then stage (8) precedes stages (2)–(7).

1.2.1 Initiation

Development is initiated when either a parcel of land (or site) is considered suitable for a different or more intensive use or demand for a particular use leads to a search for a suitable site. For the purposes of this book we will focus attention on the main uses such as shops (or retail), offices and industrial. Office and industrial uses are often combined: buildings suitable for such uses are often defined as business space. The initiative may come from any of the actors in the development process seeking an appropriate site in anticipation of the demand or need for any of the above uses. Alternatively, the initiative may stem from any actor anticipating a potentially higher value use for an existing site due to changing demographic, economic, social, physical or other circumstances. In this case, in order to identify the most appropriate use, the initiator will seek to research the market and the potential to obtain the necessary statutory planning consent for the change of use. The role of the various different actors in initiating the process will be examined later in Section 1.3 of this chapter. The initiator may not necessarily be involved in the rest of the development process, depending on their particular motive or objective.

1.2.2   Evaluation

This is the most important stage of the development process as it guides the decision making of the developer throughout. Evaluation includes market research, both in general and specific terms, and the financial appraisal of the proposal. The methods of assessing the financial viability of a project are well established (these are dealt with in Chapter 3). However, much less attention is given to detailed market research although this is continually improving (this is the subject of Chapter 7).

The process of financial evaluation needs to ensure that the cost of the development is reasonable. In the case of private sector development, this will establish the potential for profit in relation to the risk to be borne. For the public sector and non-profit-making organizations, it will attempt to ensure that the costs are recovered. An additional objective of the financial appraisal is to establish the value of the site.

This stage of the process should be carried out before any commitment is undertaken and while the developer retains flexibility. Evaluation involves the combined advice of the developer's professional team but the decision to proceed and bear the risk rests with the developer. The responsibility for this stage rests ultimately with the developer. It is a continuous process with constant monitoring, relating directly with all the other stages.

1.2.3   Acquisition

Once the decision to proceed is taken, there are many preparations to undertake before the site can be acquired and the development started. These should include the following:

(a) Legal investigation

Unless the developer is the existing site owner all legal issues concerning the site must be assessed, including ownership; existing planning permissions; and any rights-of-way, light or support. Careful preparation is required to establish who are the existing owners of all the rights to the site and what will be necessary to acquire them. Any error in establishing the extent of ownership and the cost or the time in acquiring the rights to the site may seriously affect the viability of the development. The public sector may become involved in the acquisition stage, to assemble a large site with many occupiers and landowners, as they can use their legal powers of compulsory purchase. However, the use of such powers can be time consuming and costly. The vast majority of development is undertaken through the co-operation of the original site owners, either by disposing of their interests through negotiation or by becoming partners in the development.

(b) Ground investigation

A thorough physical assessment of the capabilities of the site to accommodate the proposed use should be undertaken. This will involve the assessment of the site's load-bearing capacity, access and drainage. All existing services (electricity, water, gas and telephone) should be surveyed to ascertain their capacity to serve the proposed development. If the services are inadequate the developer needs to assess the cost of their provision or expansion. The investigation should highlight the existence of underground problems such as geological faults, made-up ground, together with the presence of any archaeological remains, contamination, underground services and storage tanks. A site survey will be undertaken to establish the measurement and configuration of the site.

(c) Finance

The developer, unless using internal resources, must also obtain appropriate finance for the development project on the most favourable terms, before commitment to a scheme. The subject of finance will be dealt with in Chapter 4. The developer will normally be concerned with arranging two sorts of finance. Firstly, short-term finance to cover costs during the development process. Secondly, long-term finance – sometimes called 'funding'– to cover the cost of holding the completed development as an investment or, alternatively, to secure a buyer for the completed scheme. The level of detailed information required by the providers of the finance vary, but all will require convincing evidence of the ability of the developer, and the soundness of the preparation and appraisal of the scheme.

1.2.4 Design and costing

Design is an almost continuous process running in parallel with the various other stages, getting progressively more detailed as the development proposal increases in certainty. The developer may have detailed knowledge of what design is required as the likely occupier is known or has been secured. In the case of a speculative scheme, the developer may need to work on a number of initial ideas with the agents and the professional team before establishing a design brief for the project. The brief is particularly important for complex schemes as it sets the design parameters for the architect.

Initially, design work will be kept to the minimum in order to keep costs down before the developer is committed to the scheme. However, there should be enough detail to enable the quantity surveyor to prepare an initial cost estimate; which the developer needs to prepare the financial evaluation. In most cases this means scaled layout plans showing the position of the proposed building(s) on the site, together with simple floor plans showing the internal arrangement of the building on each floor. Plans of the main elevations of the proposed building(s), together with an outline specification of the building materials and finishes, are often desirable. These plans together with the initial cost estimate should enable the developer to prepare the initial evaluation. By the time a decision has been made to submit a detailed planning application for the proposed scheme the initial plans will be in much greater detail. There will be a full set of plans showing the layout, elevations and section of the building, together with a detailed specification. The developer needs increasing certainty over costings to improve the quality of the financial appraisal. The quantity surveyor should be able to make a detailed estimate of the building cost at this stage to enable negotiations to commence with building contractors. Care in this preliminary work can save time and expense at later stages of the development process.

The design and costing stages involve all members of the professional team and continues throughout the construction of the scheme. The developer has to ensure that at each appropriate stage the design and costings are complete to avoid delays to the process.

In the majority of cases the final product will be very different to the initial design concept, and may go through many design changes before the final drawings are completed. The developer has to ensure where possible significant, and potentially costly, design changes are minimized when the commitment stage is reached.

1.2.5 Permission

Any development (with a few minor exceptions), which by statutory definition involves a change of use or a building operation, requires planning permission from the local planning authority prior to its commencement. The details of the planning process are dealt with in Chapter 5.

The developer may, where a building operation is involved, apply for what is termed an outline application before full approval is obtained. An outline planning consent establishes the approved use of the site and the permitted size or density of the proposed scheme. The developer only needs to provide sufficient information to describe adequately the type, size and form of the scheme. An outline planning consent, on its own, does not allow the developer to proceed with the development scheme; a further detailed planning consent is still required.

A detailed application will typically involve the submission to the planning authority of detailed drawings and information on siting, means of access, design, external appearance and landscaping. It is not possible to apply for outline consent for a change of use. There may be a number of outline applications made on a particular site if circumstances change before a developer acquires the site. If the scheme changes, after detailed consent has been obtained, then further approval is required from the local planning authority.

The developer will need to make realistic initial estimates of the likely time and cost of obtaining the appropriate permission during the evaluation stage. The acquisition of planning permission can become complex, requiring detailed knowledge of the appropriate legislation and policies, as well as local knowledge of how a particular planning authority operates. The employment of 'in-house' experts by a developer or the use of consultants may be necessary and cost-effective where planning problems are envisaged or encountered. Where permission is refused by the local planning authority the developer may appeal to the Secretary of State.

The developer may be required to enter into a contract with the local planning authority where a 'planning agreement' is negotiated as part of the planning approval. These agreements, often referred to as 'planning gains', deal with matters that cannot be covered as conditions to the planning approval, the provision and maintenance of a public facilities as part of a scheme. Planning agreements must be signed before approval is granted and often impose additional development costs therefore affecting the evaluation of the scheme.

There are a variety of other legal consents that may be necessary before a development may commence. These include listed building consent (the right to alter or demolish a 'protected' building); the diversion or closure of a right-of-way; agreements to secure the provision of the necessary services and infrastructure; and, in all cases where building operations are involved, building regulation approval. The prudent developer must clear all legal permission hurdles before commitment to the development.

1.2.6   Commitment

A developer must be satisfied that all the necessary preliminary work has been carried out before any substantial commitment is made relating to the development. Ideally all the appropriate inputs of land, finance, labour and materials, and the acquisition of statutory permissions must be satisfactorily negotiated before any agreements are signed making the developer liable for any major outlay of money. When the preliminary work has been completed as far as possible, the project must be evaluated once again. It may be that the preparation of the scheme has taken some time and the economic circumstances which determine the success of the development have changed. It is critical, therefore, that the developer pauses for thought until absolutely satisfied that the evaluation is based on the best possible information and the scheme is still viable.

Until the land is acquired, the developer must keep costs to a minimum. The likely costs up to this stage are professional fees and staff time. Depending on the circumstances some of the professional team may be willing to work on a speculative basis or at reduced fee in order to secure full appointment once the scheme commences. In some cases the developer may be acquiring the land without the benefit of planning permission and, therefore, the contract may be made subject to obtaining the necessary planning approval. In addition, conditional contracts to acquire a site are often entered into when either the developer has had insufficient time to carry out all the important preliminary investigations or the developer is yet to secure the necessary finance.

At some time, all the contracts to acquire the land, secure the finance and appoint the building contractor together with the professional team will be signed. These contracts may not necessarily be signed together: the developer must aim to achieve this as profits will be maximized. In the case of a non-profit development, ensuring that the commitment is held back until all the resources are in place will minimize cost and risk.

1.2.7   Implementation

Implementation begins once all the raw materials of the development process are in place. A commitment has now been made to a particular site and to particular buildings at a particular cost spread over a particular time. The flexibility that was possible in the earlier phases has been lost. This emphasizes once again the importance of careful evaluation and of maintaining flexibility as long as possible.

The aim throughout this stage is to ensure that the development is completed within the time and budget set out in the evaluation, without comprising quality. Depending on the experience of the developer and the complexity of the scheme, this may best be achieved by employing a project manager to co-ordinate the design and building process. The project manager and/or developer must anticipate problems and make prompt informed decisions to minimize delays and extra costs. The developer must take as much interest in the running of the project as in its promotion. The market must be constantly monitored to ensure that the product is right, which, may involve changing the specification. In the case of a non-profit development, the developer must aim to contain costs, while maximizing the benefits of occupation.

The construction and project management stage of the development process is dealt with in Chapter 6.

1.2.8   Let/manage/dispose

This phase of development, though often the last stage, will need to be at the forefront of the developer's thoughts from the initiation of the scheme. In some cases the occupier may have been secured at the start or during the development process. The development's success will depend on the ability to secure a willing occupier at the estimated rent or price, within the period originally forecast in the evaluation. The disposal may take the form of a letting or it may be the outright sale of the freehold interest. In the case of a major retail development there are many lettings, while in that of an single office building the property may be disposed of in one major letting.

The letting and/or sales strategy should be thought out during the evaluation stage and updated, where possible and appropriate, during the course of the development. Any agent or a member of staff employed by the developer to secure lettings/sales should be included in the development from the start of the process. A decision must be made at what point it would be sensible to let or sell. In many cases it is necessary to complete or virtually complete the development before seeking an occupier. This decision may not be the developer's alone and may be heavily influenced by other actors in the process such as the financiers or the landowner (if they have remained a partner in the development).

At the start of the process the developer has to decide whether the property investment created is to be held as such or sold to realize any profit (unless it has been pre-sold to the long-term financier of the scheme). This will depend on the motivation of the developer and the prevailing property investment market conditions at the time. However, developers have to be flexible to accommodate any changes in the investment market prior to completion of the scheme. Careful thought needs to be given to the investment value at the initial evaluation and design stages. If the decision is made to sell the investment to an investor then the developer needs to research their requirements. The location, specification and financial strength of the tenant(s) will be critical in achieving the best price for the investment. The developer may employ an agent to secure a sale of the property to an investor. The agent should be employed as early as possible to advise on the specification and design of the scheme.

The development process and the developer's responsibility should not end with the occupation of the building. There is still a need for the developer to maintain contact with the occupier, even though no direct landlord/tenant relationship may exist. Developers can learn more about occupiers' requirements in general and, in particular, the shortcomings of the completed building from a management point of view. Management needs to be considered as part of the design process at an early stage if the final product is to benefit the occupier and earn the developer a good reputation. The financial success of the development cannot be assessed until the building is complete, let and, where appropriate, sold. Often it may not be until the first rent review under the terms of the letting (typically 5 years after occupation), that the picture will become clear.

1.3 MAIN ACTORS

The development process has been reviewed and divided into stages. Within each stage, and across some or all of them, there are a variety of important actors, who each contributes to the outcome of the process and who may have very different perspectives and expectations. In a lecture at Reading University in 1983 Ron Spinney (currently chief executive of Hammerson plc) compared the role of the developer with that of a director of a play who has to manage the diverse and conflicting objectives of all actors on a public stage:

> Any performance must have a director who has the capacity and energy to pull together the performance to ensure it reaches a satisfactory solution.

The actors are considered below in approximately the order they appear in the development process. Their importance varies from project to project and not all of them appear in every development scheme.

1.3.1 Landowners

Landowners play an important role, whether actively or passively. They may actively initiate development by a desire to sell and/or improve the value of their land. If they are not the active initiators, they may be a crucial hurdle to the development, for without their willingness to sell their interest or participate in the development (unless compulsory purchase powers are used), no development can take place. The landowner's motivation may well affect their decision to release land for development, whether they be individuals, corporations, public authorities or charities. They may even take on the role of the developer, either in whole or in part. Massey and Catalano (1980), in their study of land ownership, categorize ownership into three broad types which are paraphrased here into the traditional, industrial and financial.

Traditional landowners include the church, landed aristocracy and gentry, and the Crown Estate. These are significant owners in terms of area and capital value. They are distinguished from other categories by not being entirely motivated by the economic imperative. The motives for ownership are wider than return on capital and involve social, political and ideological constraints.

Industrial landowners own land incidental to their main purpose, which is some form of production or service provision. This category includes a wide variety of types, including farmers, manufacturers, industrialists, extractive industries, retailers and a variety of service industries. Public authorities of all types, such as central, local and nationalized industries who own land incidental to providing a service or product, might also be included in this category. The motives of this group are complex, in that they are both dominated and constrained in their attitude to land by their main reason for existence – their product. Also they may be constrained by their legal status and will not always be seeking to maximize their return on land or property in the narrow sense since that would be seen as subservient to their main purpose. Thus for them, the economic advantages of releasing land for development are not always evident. If they are forced to sell their land due to a compulsory purchase order, then although they are compensated for the costs of relocating and disturbance to their business, no allowance is made for the fact they are unwilling sellers (unless they are residential occupiers). There are intangible losses to commercial businesses which are difficult to value.

Financial landowners see their ownership as an investment like any other and may therefore be expected to co-operate with development if the return on their land is financially optimal. This category of owners has obvious financial motives and are likely to be the most informed type of owner regarding land values and the development process. The major group in this category are financial institutions (pension funds and insurance companies) who own a significant proportion of land by capital value and have been investing on average between 5 and 15% of their considerable funds in property investment over the last 20 years. They may also act as developers directly or in partnership with developers. Also included are the major property companies who own substantial portfolios of properties and carry out development, and who may therefore act as both landowner and developer (see below).

In the past landowners have had an important impact on both the spatial layout and the type of development constructed, e.g. the Grosvenor and Bedford estates in London. The planning system has lessened the impact they are able to have on the type of development, but they can still influence the location and planning of the development. The number of owners involved in any particular development also has an important effect, as the greater the number of owners and the smaller their holdings, the more difficult it is to assemble a development site. Many developments have taken years to come to fruition, requiring great patience on the part of the developers.

1.3.2   Developers

Private sector development companies come in a variety of forms and sizes from one-man-bands to multinationals. Their purpose is usually clear: to make a direct financial profit from the process of development – in the same way that any other private sector company operates, whatever their product. Developers either operate primarily as traders or investors. Most small companies have to trade, that is to sell the properties they develop, as they do not have the capital resources to be able to retain their completed schemes. Many larger public quoted development companies in the 1980s (often referred to as merchant developers) preferred to trade developments to capitalize on rising rents and values. Unfortunately they borrowed money on the strength of their future profits and the majority went into receivership during the crash of the early 1990s as their limited assets were insufficient to support them. Some survive but are effectively controlled by their bankers. Many trader-developers try to evolve into investor-developers as success enables them to retain profits for investment purposes. At the other end of the scale, some of the largest companies – in terms of capital assets – do hardly any new development at all, being content to manage their property portfolio and undertake only refurbishment and redevelopment work. Residential developers operate almost solely as traders as the market is heavily biased towards owner occupation, but many become significant landowners during the development process.

   The kind of development undertaken varies considerably. Some companies specialize in a particular type of development, e.g. offices or retail, and also in particular geographical locations; others prefer to spread their risk across types and locations. Some remain in a specialist type of development but cover a wide geographical, even international area. Property companies formulate their policy according to the interest and expertise of their directors and their perception of the prevailing market conditions.

1.3.3   Public sector and government agencies

Due to central government policy the public sector is currently undertaking little direct development. Local authorities are primarily involved with developments for their own occupation or community use and the provision of infrastructure. Local authorities are both constrained by their financial resources and limited by their legal powers. Local authorities have to be publicly accountable and have regard to the overall needs of the community they serve. Local authority involvement in the development process will depend on whether they wish to encourage development or control development in order to maintain standards. Many local authorities undertake economic development activities, with the limited resources they have to promote development and investment in their area. Furthermore many of the more active authorities act as a catalyst to the development process by supplying land, and where possible buildings, to achieve economic development of their area. Participation may be limited to the role of landowner, by maintaining a long-term interest in a development. Local authorities will often retain the freehold of their development sites and grant a long leasehold interest to the developer, sharing in rental growth through the ground rent.

Government policy has been formulated over the last 15 years on the basis of only intervening in the development process where private market forces fail to bring forward development, particularly in areas targeted for economic development. The government's urban regeneration initiatives are administered through several government agencies including Urban Development Corporations (UDCs), English Partnerships, the Welsh Development Agency and Scottish Enterprise. Their role is seen as an enabling role, bringing forward development and attracting investment, in partnership with the private sector. They are able to assist developers with land assembly, site reclamation, the provision of infrastructure, financial grants and, more recently, following the relaxation of Treasury rules, rental guarantees. Other government initiatives aimed at attracting occupiers with financial incentives to specific areas of the country include Enterprise Zones and Regional Selective Assistance. All of the various government initiatives on urban regeneration are dealt with in Chapter 2.

1.3.4 Planners

The planning system has existed in a comprehensive form since 1947 and is firmly established as the major regulator of property development (see Chapter 5 for a detailed account).

Planners can be divided into two broad categories: politicians and professionals. The politicians, usually on the advice of their professional employees, are responsible for approving the development plans drawn up by professionals in accordance with the policy laid down by themselves. They are also responsible for determining whether applications for permission for development proposals should be approved or refused. The professionals are responsible for advising the politicians and administering the system.

The main purpose of planning is to 'encourage development' and to prevent 'undesirable development'. The basis for determining planning applications is laid down by statute and a variety of central government policy guidance notes. Local government must adhere to these and determine its own local policy through the main medium of development plans. Individual planning applications are determined in the light of these development plans, written government policy and advice, previous decisions and the particular nature of the application. In practice there are a lot of gaps and conflicts in the guidance which means developers often employ planning consultants to assist them in negotiations with the planners. Developers need to know what use, what density, what design standards are required in order to obtain permission. Successful applications are usually best achieved by prior negotiation with the authorities and this may involve agreement by the developer to provide infrastructure or community facilities in the case of large developments, known as planning gain. This type of agreement is endorsed by government guidance provided it reasonably relates to the development proposed. In the context of tight public spending controls 'planning gain' is seen by local authorities as a means of securing useful benefits for the community. However, the issue of planning gain has been controversial and there is a limit as to how much can a developer can afford.

Planning authorities differ widely in their policies towards development. Those in areas of low economic activity typically wish to encourage development activity, putting only minimal restrictions on proposals, particularly those which that will provide employment. Authorities in areas of high economic activity mainly see their role as imposing higher standards, slowing down development in order to achieve a better balance of uses and improved design of buildings. There is an increasing level of conflict between developers and planners, leading to increasing use of the appeals system, especially in areas of high economic activity. In some instances the conflict is caused by the politicians ignoring the advice of the professionals.

1.3.5 Financial institutions

Unless a development is being financed entirely with a developer's own capital or that of a partner, then financial institutions, as providers of finance, have a very important role in the development process. Financial institutions is a term usually used to describe pension funds and insurance companies. However, there are many other financial intermediaries such as clearing and merchant banks (both UK and foreign), and building societies who also provide finance for property development.

Two main types of money are needed for development: short-term money or 'development finance' to cover the costs during the development process; and the long-term money or 'funding' to cover the cost of holding the completed development as an investment. Alternatively, the developer will in the long term seek a buyer for the completed scheme to repay the short-term loan and realize any profit.

Financial institutions (pension funds and insurance companies) are motivated by direct financial gain. However, unlike developers, they take a long-term view, needing to achieve capital growth to meet their payment obligations in real terms to pensioners and policy holders. Although pension, life and investment funds are judged on their short-term performance both in relation to other forms of investment and to the returns they achieve against competing funds. They seek to minimize risk and maximize future yields. The yield on any investment is the annual income received from the asset expressed as a percentage of its capital cost or value. Property is only one of a number of investments the institutions invest in and may represent only 5–15% of their entire portfolio of investments. In the case of property the financial institution will receive a lower initial income when compared to a fixed-interest investment, but this will be more than compensated by the long-term growth.

They may provide both short-term and long-term finance to a developer by what is called 'forward-funding' a development: they agree to purchase the development on completion whilst providing all the finance in the interim. Almost all the risk passes to the developer who will in the majority of cases provide a financial guarantee. Alternatively, they may act as developer themselves to create an investment: all the risk is theirs but they do not have to provide a profit to the developer. Some only purchase completed and let developments as they perceive development as too risky. In order to be persuaded to take on the risks associated with development, rather than purchasing a completed and let scheme, they need a higher return (yield).

Whether acting as developer, financier or investor they tend to adopt rigid and conservative policies, although they all differ in their individual criteria. However, they all tend to seek a balanced portfolio of property types, rather than specializing in one particular use. Most try to spread their investments geographically. They will seek properties or developments which fit their specific criteria in terms of location, quality of building and tenant covenant (financial strength). As a result developers have tended to develop schemes in accordance with the financial institutions' specification rather than occupiers. The financial institutions wish to purchase a building which will have the widest tenant appeal and therefore, their advisers usually take a conservative view and recommend the highest specification.

The developer may approach the banking sector for funding if a development is either not 'institutionally acceptable' or the developer is not prepared or unable to provide the necessary guarantees. Alternatively, the developer may prefer to use debt finance in a period of rising rents and values (as happened in the late 1980s) to maximize the potential profit on completion. There are a great variety of methods of obtaining finance from the banks both for short- and medium-term finance (these will be discussed in detail in Chapter 4). The banks also aim to make a financial profit from the business of lending money. Bank lending may take the form of 'corporate' lending to the development company or lending against a particular development project. The banks will use the property assets of the company or the property as security for the loan. Property is attractive as security as it is a large identifiable asset with a resale value. Banks wish to ensure that the proposed development is well located, the developer has the ability to complete the project and the scheme is viable. With corporate lending, the bank is concerned with the strength of the company, its assets, profits and cash-flow. Depending on the size of the loan, and in particular where the bank is exposed to above normal risk, then the banks may secure an equity stake in the scheme.

Residential developers, who build housing for owner-occupation, only require short-term development finance, usually provided by the banks. Their ability to raise finance is based on their 'track record' and the value of the scheme.

For public sector development, the sources are similar but much more difficult to obtain, with tight central government control on public sector borrowing. Some local authorities may obtain funding through grants for urban regeneration projects in specific geographic areas, mainly from central government sources and European Structural funds. However, access to such funding is subject to competition, and is targeted at schemes carried out in partnership with the private sector and the community.

Developers may also obtain financial assistance from the various government agencies in the form of grants, rental guarantees and equity participation through the provision of land. However, the developer has to prove that the project would not proceed without such assistance and that it will create jobs relevant to the local population.

1.3.6   Building contractors

Building contractors are employed by developers to construct the development scheme and their prime objective is direct financial gain. There are a great many building contractors and a considerable variety of contractual systems for obtaining a completed building (see Chapter 6 for a detailed account).

Some development companies employ their own contractors. Residential developers tend to employ all the necessary expertise in-house: this is why they are often referred to as house-builders. Development companies may keep their contracting division at 'arm's length' as an entirely separate profit-making centre. A builder who takes on the role of developer, e.g. house-builders, also takes on the additional risk associated with the development process. When a builder is merely employed as a contractor the financial profit is related to the building cost and length of contract. Under a design and build type contract a contractor will take on a design role which will involve a greater element of risk in relation to the responsibility for cost increases. Larger contractors with the relevant expertise may take on the role of a management contractor and manage all the various sub-contracts for the developer in return for a fee. In the case where the builder is the developer then a larger return is required due to the risk involved but by combining the building and development profit, an overall lower profit is acceptable. For builders who employ a substantial labour force, an additional objective may be to ensure continuing employment. Sometimes this can only be achieved by cutting tender prices and therefore profits.

Essentially, building contractors carry out a specialist activity within the development process, commencing at a time of maximum commitment and risk for the developer. The prudent developer will therefore ensure the capability and capacity of the contractor(s) to undertake the proposed work, seeking the right balance between the lowest tender and quality of performance. However, it is not in the contractor's or developer's interest to have a situation where the contractor cannot make a reasonable profit from the scheme. It is not in the developer's interest for the contractor to go bankrupt or to be compromising on quality.

1.3.7   Agents

Commercial agents or estate agents (in the case of residential) may be instrumental in initiating the development process and/or bringing together some of the main actors in the process. They also form the link between the developer and the occupier, unless the developer uses 'in-house' staff to perform the agent's role and the occupier is not represented by an agent. Therefore, they play a very important role within the development process and they are often involved in every stage of the process. Agents are able to perform this role due to their detailed knowledge of both the property market in terms of demand and current rents/prices and their 'personal' contacts with developers, occupiers, financial institutions and landowners. This emphasizes the fact that the development industry is all about 'people'.

Their aim is to make a direct financial profit from the fees charged to their client (be it the developer or occupier) for carrying out a professional service. In the case of introducing one party to another they only receive a fee if the transaction is completed, e.g. the property is let, but it is nearly always related to the value of the transaction in percentage terms.

They may be instrumental in initiating the development process by either finding a suitable site for a developer or advising a landowner to sell a particular site due to its development potential. Unless they are retained by a developer to specifically find sites to suit a particular use, they take the initiative in identifying suitable sites and 'introduce' them to developers.

They will introduce sites to those developers who they consider have the appropriate expertise and resources to both acquire the site and complete the development. The agent will negotiate with the landowner on the developer's behalf and advise the developer on all matters relating to the evaluation stage. In return they may secure not only a fee for introducing the site (usually related to the purchase price), if the acquisition stage proceeds, but also they may secure appointment as letting and/or funding agent for the development scheme. It can be a lengthy and time consuming process for the agent but the rewards can be high. If an agent acts for a landowner then they advise on both the likely achievable land value and the likely market for the site.

Agents may be used by developers to assist them in securing the necessary finance for a development scheme due to their knowledge of the requirements of the financial institutions or banks. Many institutions retain agents to advise them generally on their property investments including development funding: they may specifically find development opportunities for their client to fund. The institution's agent will normally advise their client throughout the process and be one of the letting agents on the scheme. Some of the larger or more specialist firms of chartered surveyors, with financial service divisions, may act as financial intermediaries arranging funding packages with banks and other institutions in return for a fee related to the size of the loan.

Agents are widely employed by developers as letting or selling agents, providing that all important link between developer and occupier. In performing this role they should be involved in the development from the start to enable them to advise the developer on the occupier's viewpoint. However, unless they have a specialist marketing department they usually cannot give total comprehensive advice on the precise requirements of the market. A developer will need to commission market research to obtain more detailed knowledge of the specific market for the completed development. Some developers may employ an in-house team but the advantage of the agent is their knowledge of the market in general and their contacts with potential occupiers or their agents.

Developers, landowners, occupiers, financiers and property investors may at some stage in the development process employ a suitably experienced qualified chartered surveyor or valuer to undertake professional work to assist their decision-making process. Chartered surveyors and valuers are employed by the vast majority of commercial and residential agents to enable them to undertake professional work alongside their agency work. Professional work related to development may include valuation, building surveys and management. Developers will require an independent valuation to check their own opinion of value, particularly where they have insufficient knowledge themselves or it is required by a financier of the development. Independent and in-house valuers are also used by financial institutions and banks on schemes for which they are considering making loans or granting mortgages, including the asset value of any security being provided by developer. Financiers will often employ building surveyors to check on the construction phase of the development to ensure that it is being built to the right specification and to certify drawdowns of the development loan. In the public sector local authorities, central government and the Inland Revenue all use valuers to advise and check on any development related work.

1.3.8   Professional team

Most developers do not have the skills or expertise to carry out a major development: the development process is complex. Consequently, developers employ professionals to advise them at various stages of the development process; these include the following.

(a) Planning consultants

Planning consultants are employed to negotiate with local planning authorities to obtain the most valuable permission for a development, particularly with large or sensitive schemes. If a planning application is refused they will be employed to act as expert witnesses in presenting the case for the developer. They also advise landowners to ensure that the sites within their ownership are allocated within the development plan to their most appropriate or valuable use. This may involve negotiations with the local planning authority at plan preparation stage or subsequent representations at an enquiry into the development plan. In performing this role they can be important initiators of the development process.

(b) Economic consultants

Economic consultants are employed at the all important evaluation stage to provide a detailed analysis of the characteristics of the market in terms of the underlying demand and competitive supply. Many financiers, particularly the financial institutions, insist on market analysis when considering a development funding proposal. In addition, they often employ researchers in-house in formulating their funding criteria and policy (see Chapter 7 for further details on the use of research).

(c) Architects

Architects are employed by developers to design the appearance and construction of new buildings or the refurbishment of existing buildings. They may also administer the building contract on behalf of the developer and certify completion of the building work. In the case of refurbishment work, building surveyors are usually employed to survey the existing building and advise on the alterations. Architects are normally responsible for obtaining planning permission where a planning consultant is not employed. They are paid on a fee basis, usually a percentage of the total building contract sum.

It is important that the architect is employed at the earliest possible stage to ensure that all design work is ready when construction is due to commence. It is also important to employ architects with the appropriate experience, reputation, resources and track record. A developer should ensure the architect has the right balance of skills to produce not only good architecture but a cost-effective and workable design attractive to occupiers. This balance is very often hard to achieve and it is therefore important for a developer to produce a clear architectural brief from the start. Problems start when there is insufficient communication between architect and client.

Some firms of architects offer a comprehensive service, including project management, engineering and interior design work. This may be effective on some development schemes but most developers tend to prefer to assemble their own professional teams to achieve the most effective and experienced team. Some development companies, particularly those which specialize, employ in-house architects and design professionals.

(d) Quantity surveyors

Quantity Surveyors are in simple terms 'building accountants' who advise the developer on the likely costs of the total building contract and associated costs. Their role includes costing the designs produced by the architect, administering the building contract tender, advising on the most appropriate form of building contract, monitoring the construction and approving stage payments to the contractor. They are increasingly becoming involved in the administration and management of design and build contracts. Like architects, their fee is based on a percentage of the final contract sum. The choice of quantity surveyor should be based on appropriate experience and reputation. The developer should select a quantity surveyor who works well in partnership with architects and other members of the professional team to produce cost-effective designs. A quantity surveyor should be able to provide the developer with cost-effective ideas as alternatives to those proposed by the architect.

(e) Engineers

Structural engineers are employed to work with the architect and quantity surveyor to advise on the design of the structural elements of the building. They will also participate in the supervision of the construction of the structure. Civil engineers will be employed where major infrastructure works and/or ground work is required. On large and complex schemes mechanical and electrical engineers are used to design all the services within the building. Engineers are usually paid a percentage fee based on the value of their element of the building contract.

(f) Project managers

Project managers are employed to manage the professional team and the building contract on behalf of the developer. Project managers are normally only employed on the larger and more complicated schemes.

Developers often act as project manager or rely on in-house staff or another member of the professional team. They should be appointed before any of the other professional team or the contractor so that they are in a position to advise the developer on the best professional team for the project. Their fees are normally based on a percentage of the building contract sum, but often they may be an incentive for managing the scheme within budget and on time.

Developers often perform the role of project managers themselves for occupiers who wish to employ the expertise of a developer in constructing their own premises.

(g) Solicitors

Solicitors are needed at various stages throughout the development process, starting with the acquisition of the development site through to the completion of leases and contracts of sale. They are often involved with the legal agreements covering the funding arrangements entered into by the developer. If the developer has to appeal on a planning application then both solicitors and barristers may be involved in presenting the developer's case at an enquiry.

(h) Accountants

Specialist accountants may be employed to advise on the complexity of tax and VAT regulations which can have a major cost impact on a development. They may also provide advice on the structure of partnership or financing arrangements.

The above is not intended as an exhaustive list of the various professionals and specialists employed during the development process. There are a considerable variety of other specialists who may be necessary depending on the circumstances of the project and its complexity. Others may include highway engineers, landscape architects, land surveyors, soil specialists, archaeologists, public relations consultants and marketing consultants. The above shows the variety of skills that are required within the development process.

1.3.9   Objectors

There are two categories of objectors who can potentially cause delay and possible abandonment of development projects. The first may be purely 'amateurs' and self-interested neighbours of the proposed development.

They are often referred to as 'NIMBYS' ('not in my back yard') and, where organized, they can achieve considerable obstruction to the progress of development proposals. In the 1980s such well-organized and influential organizations contributed towards the abandonment of proposals for private sector New Towns in Oxfordshire and Hampshire.

The second category are the well-organized professional, permanent bodies at local and national level. Locally they include amenity societies who take an interest in every proposal affecting their local environment and heritage. Nationally they include the Victorian and Georgian societies. These bodies have considerable influence with the local planning authorities and tend to be always consulted on major applications. There are also official quangos (quasi non-government organizations), such as the Royal Art Commission, the Nature Conservancy Council and English Heritage, who may take an interest in important buildings and existing flora and fauna. These organizations are well informed, and have a good knowledge of the planning and development processes.

Developers must be aware of their interest (or likely interest) and be prepared to accommodate or refute their opposition. Ideally such negotiations should be carried out before a planning application is made to avoid lengthy delays. Opposition can be costly to a developer, either by imposing higher standards and costly alterations or lengthy delays. At worst opposition can lead to the complete abandonment of proposals, which may have been unsound. Prudent developers need to account for objectors when evaluating their proposals.

1.3.10 Occupiers

Unless the occupier of a building is the developer or is known early in the process, then the occupier is not regarded as a main actor within the development process, as they are often unknown until the development is complete and let/sold. Their demand for accommodation triggers the development process and influences both land prices and rents, to which developers respond (see Section 1.4). The occupier is an actor within the process and their requirements should be researched at the beginning of the process. Developers in the past have tended to produce buildings in accordance with the requirements of the financial institutions and the needs of the occupier have been overlooked. However, there is a growing recognition within the industry that this has to change and many developers now commission research into occupiers' requirements at a general and specific level (see Chapter 7).

When the occupier of the scheme is known early on in the process, then the occupier becomes the most significant actor in the process. The building will be constructed in accordance with the occupier's requirements, which can be very specialized, particularly with industrial users. The developer may need to persuade the occupier to compromise on their requirements to provide a more standard and flexible type of building, so that the investment market for the building is wider in the event of disposal. The developer will also be concerned to protect the value of the building as security for loan purposes.

Occupiers, as commercial businesses, largely tend to regard the buildings they occupy as an overhead incidental to their activities as providers of a service or product. Although some companies do employ an in-house property team (consisting of surveyors and facilities managers) and many are set up as a profit-making centre, in some instances acting as consultants to other occupiers (e.g. Digital's facilities management consultancy), many occupiers tend to fail to adequately plan their property requirements far enough in advance, reacting to changes in their business as they happen. The property requirements of occupiers are influenced by both the short-term business cycle and long-term structural changes underlying the general economy (see Section 1.4 and Chapter 7). Either of these factors may influence occupiers at a specific level or across the business sector in which they operate. Their demand for accommodation is also influenced by advances in technology affecting both working practices and their physical property requirements.

Both agents and developers criticize occupiers for not knowing what they want, although many companies are becoming far more knowledgeable about the role of property within their businesses and their requirements in terms of specification. This is evidenced by the growing profession of facility management. Occupiers have differing requirements and priorities, particularly in the case of offices, making the task of the developer difficult in producing a building suitable for as many tenants as possible. The response of the financial institutions is to seek the highest quality specification with a layout to suit the widest possible range of tenants. As a consequence an occupier may be forced to occupy a building which compromises their requirements in terms of location or specification. There were many examples in the late 1980s of occupiers stripping out buildings and refitting them in accordance with their requirements. It is impossible to achieve a multi-purpose building as clearly shown in the City of London, where financial service companies require large open plan floors and high ceilings, and professional services require cellular offices with good natural daylight.

Another area of conflict between occupiers and the financial institutions concerns lease terms. Occupiers are increasingly demanding more flexible lease terms to enable them to respond in the short-term to their property requirements. Financial institutions prefer the traditional 25 year lease term with upward-only rent reviews. However, many will accept shorter lease terms (from 15 years) and options to break. Tenant pressure groups are trying to persuade the government to legislate against some of the provisions of institutional leases (upward-only rent reviews and privity of contract) but have so far been unsuccessful. Instead, a code of practice is being drawn up by organizations representing both landlord and tenant organizations.

There is evidence that developers and the financial institutions are taking more account of the needs of occupiers. There is a recognition that the buildings of the 1980s were over-specified and that reducing the costs of occupying a building are important to occupiers. The current trend is towards energy efficient buildings with the maximum use of natural ventilation and the use of air-conditioning where absolutely essential.

1.4   ECONOMIC CONTEXT

Property development does not, of course, occur in a vacuum. Occupier demand is a reflection of the short- and long-term changes in the economy. Availability of development finance is also linked to conditions in the wider economy. The economic context is important to developers in both a specific way (in so far as the local economic context helps to determine the market for a individual scheme) and in a more general way via the wider economy (as it affects general property market conditions and the confidence of occupiers, investors and developers).

1.4.1   The local economy

Research studies have shown that most demand for an individual office or industrial development is drawn from a small geographical area around the scheme. Similarly, the ability of a retail scheme to attract national retailers depends crucially on the spending capacity of the local population. Since local economic conditions will help determine how much development, and of what type, is appropriate in a particular place, it is in the interest of any developer to look beyond the individual scheme to the wider economy. The local economic context can thus be a useful indicator of the likely viability of any development project, and can be used alongside the development appraisal. The recent *Mallinson Report* on property valuation also urges greater appreciation of local economic conditions in valuation advice (Royal Institution of Chartered Surveyors, 1994). The role of market research is covered more in Chapter 7.

1.4.2   The national economy

At a more general level, the interaction of the short-run business cycle with property cycles creates great variability in a developer's plans, and the ability to progress schemes at different times. The developer also needs to be responsive to the more evolutionary changes which occur in occupier preferences as a result of long-term changes in the structure of the economy.

(a) The business cycle

Research in the last decade or so has established the nature of the link between the economic, or business, cycle and the property market. Useful references on this complex topic include various papers on building cycles by Richard Barras (1994) and the recent Royal Institution of Chartered Surveyors report on property cycles, carried out by researchers from the Investment Property Databank and Aberdeen University in 1994.

Three important cycles have been identified, all of which exhibit different periodicity – the business cycle (which drives the occupier market), the credit cycle (which influences bank and institutional funding) and the property development cycle itself. A simplified version of Barras' analysis follows beginning with the business upturn:

1. Strengthening demand, rising rents and capital values trigger the start of the new development cycle upswing.
2. If credit expansion accompanies the business cycle upswing, it can lead to a full blown economic boom. The banks may also fund a second wave of speculative development activity.
3. However, because of the long lead times in bringing forward new development, supply remains fairly tight and values continue to rise.
4. By the time the development cycle reaches its peak, the business cycle has already moved into a down-swing, accompanied by a tightening of monetary policy to combat the inflationary effects of the economic boom.
5. As the economy subsides, the demand for property declines; rents and values fall as a result and the vacancy stock increases in supply.

6. As the economy moves into recession, the fall in rents and values continues, property companies are hit by the credit squeeze, bankruptcies increase and the development cycle is choked off.

As Barras also points out, the experience in Britain has been that a property boom of the scale of the late 1980s is typically followed by a more muted development cycle. Banks and investors are struggling with debts incurred during the most recent recession and are disinclined to fund speculative development. At the same time, oversupply of property built in the previous boom is sufficient to meet demand during the whole of the following business cycle. It is only when supply is exhausted, at the start of the next business cycle upswing, that the speculative development cycle takes off once more. Clearly, the likely success of development projects will be influenced by where in the cycle they are started and completed.

(b) Structural change

Underlying the short-term business cycle are longer-term shifts in occupier requirements which result from structural changes in the economy. For example, the recent expansion of demand for very large warehouses has resulted from strategic reorganization within the retail and logistics industries, helped by the increasing availability of sophisticated information technology. Similarly, the changes in working practices amongst office occupiers, again encouraged by developments in information technology, will most likely generate demand for new kinds of office building in the future. Developers (and investors) who monitor these long-term changes can begin to create new types of product ahead of the rest of the market; equally they can avoid being left with buildings which have a diminishing 'shelf life'. This is one area in which property market research can be of great use, a subject covered more fully in Chapter 7.

Executive Summary

In this chapter we have taken an overview of the very complex activity of property development, reviewing it as a series of stages involving many actors with differing objectives, operating within the overall context of the building cycle and its interaction with the business and credit cycles. Development may be initiated by any of the main actors identified but it can only take place with the consent of the landowner (unless compulsory purchase powers are used). As a development proceeds through the various stages the developer will become increasingly committed and flexibility will be reduced, exposing the developer to greater risk. Before developers make a commitment to both acquiring land and signing a building contract they should obtain all the necessary consents, carry out all the necessary investigations and secure the necessary finance. In addition, a thorough financial and market evaluation should be carried out with the best information possible to establish the project's viability and the occupier market. The success of a development may often depend on luck rather than the developer's judgement and skill, depending on the interaction of the building and business cycle on completion.

REFERENCES

Barras, R. (1994) Property and the economic cycle: building cycles revisited. *Journal of Property Research*, 11, 183–197.

Barrett, S. *et al.* (1978) *The Land Market and Development Process*, School for Advanced Urban Studies, University of Bristol.

Goodchild, B. and Munton, R. (1985) *Development and the Landowner*, Allen and Unwin, London.

Massey, D. and Catalano, A. (1980) *Capital and Land*, Edward Arnold, London.

Royal Institution of Chartered Surveyors (1994) *The Mallinson Report. Report of the President's Working Party on Commercial Property Valuations*, Royal Institution of Chartered Surveyors, London.

University of Aberdeen and Investment Property Databank (1994) *Understanding the Property Cycle*, Royal Institution of Chartered Surveyors, London.

2

LAND FOR
DEVELOPMENT

2.1 INTRODUCTION

The acquisition of land is usually the developer's first major commitment to a development project. The initiation, evaluation and acquisition stages of a development are closely linked and very often run in parallel. However, the acquisition of the land should not be completed until after the evaluation process (see Chapter 3). This chapter will look at both the initiation and acquisition stages of the development process.

The selection of a site fundamentally affects the nature and success of a development. No amount of careful design or promotion can totally overcome the disadvantage of a poor location or a lack of demand for the accommodation at an economic price irrespective of location. Land is unique and every site has its own characteristics. The site acquisition process can be very frustrating and unpredictable as many factors, some outside the developer's control, influence its success.

This chapter will examine the methods whereby developers identify and acquire sites. The role of landowners as initiators of the process will be discussed. It will also examine the role of both local authorities and the government's urban regeneration initiatives in bringing forward land for development and attracting private sector investment. It will include an examination of the grants and tax incentives available to the developer in areas of the country where the implementation of development schemes might not take place if left to market forces.

2.2 SITE IDENTIFICATION

The first step in finding a development site is to establish a strategy defining the aims, nature and area of search. The starting point may be the development company's own business plan which sets out the aims of the company. As already discussed in Chapter 1, development companies may restrict themselves geographically or to particular types of development, e.g. out-of-town retailing. The overall strategy and aims of a development company will form the basis for the identification of sites and development opportunities. Within this overall strategy a developer needs to define more closely specific areas of search and their exact requirements, in relation to the size, nature and location of sites. The geographical area of search for sites depends on a number of factors. This may depend on the location of the developer's own office or offices, the desire to spread risk across a number of locations, the availability of development finance and the results of market research.

The location of the company's offices is an important criterion as the further the site is away from the office the more likely that the management of the project will be less effective. If the developer is operating within a local area then the developer will tend to establish good contacts with agents, occupiers and the local planning authorities. In the case where a development site is a considerable distance from the developer's office a good working relationship will need to be established with local agents and it may be prudent to carry out the development with a local partner. There is no substitute for local knowledge. However, development companies, particularly the larger public quoted ones, may wish to spread their exposure to risk by spreading their development activities over a number of different locations, so if an oversupply of accommodation occurs in one location it only affects some projects and not the entire development portfolio.

A developer's plans for the identification of sites may be largely influenced by the way in which the developer usually finances development projects. If a developer intends to seek finance from a financial institution then the developer needs to be aware of their preferred locations for property investment. The requirements of financial institutions will be discussed in greater detail in Chapter 4. During the late 1980s development boom, the location of speculative development sites became less critical for the purposes of obtaining finance, with the widespread availability of bank finance and the decline in the dominance of the financial institutions as funders in the development market. In any market conditions the obtaining of development funding from either banks or financial institutions is likely to be more successful on well-located 'prime' sites, as the schemes built on such sites will have the widest tenant appeal and hence reduce risk. Whatever the state of the property market the prudent developer should always seek the best locations appropriate to the proposed use. A secondary location will represent a higher risk to the developer in terms of achieving a letting/sale and securing funding which should be recognized in the evaluation of the scheme (see Chapter 3).

However, the developer's perception of occupier demand, backed up by market research, should be the most important factor in influencing the area of geographical search. The developer's skill and knowledge is important in identifying areas of potential growth where market forces will provide demand for accommodation which will exceed supply by the time a development project is on stream. Forward-thinking developers may commission research at a strategic level to identify trends in the market and areas of potential opportunity (see Chapter 7 for further details on strategic research). Whilst recognizing the risks involved, the developer will always seek opportunities to be ahead of the market. Road and public transport improvements together with major redevelopment schemes are some of the more obvious factors which may influence the demand for accommodation in a particular area. Developers will seek to purchase a site in a town which is likely to benefit from a planned major road scheme. For example, many developers bought sites in towns on London's M25 orbital motorway as its route became known in the early 1980s. Market research should seek to identify the current and projected levels of supply and demand of various types of accommodation in a particular area. The developer should also examine trends in rental and capital values, and the underlying economy in the areas of search identified. The use of market research in formulating a land acquisition strategy is discussed in greater detail in Chapter 7.

As part of any market research the developer should identify the factors which influence occupiers in their choice of location. We will now examine the various factors which influence the choice of location for each type of commercial development.

(a) Office/'business space' development

The majority of office development tends to take place in London and the South East. Outside the South East, office development activity is concentrated in the major cities and regions, e.g. Bristol, Birmingham, Manchester, Leeds, Edinburgh and Glasgow. According to research by the surveyors Chesterton, in 1992, 8.6 million sq ft of office space was completed, of which the majority was in London.

Road, rail and air communications are vital in the consideration of locations for office development. In London and Britain's other main cities the proximity to good public transport is important for office locations in the central areas. In relation to office development in provincial towns and 'out-of-town' business or office parks, proximity to the national motorway network and airports are important. This has been demonstrated in the South East with the growth of towns along the motorways (M3, M4 and M25) and near London's airports (Heathrow and Gatwick).

At a specific level the locational choice of an office occupier is determined by such diverse factors as tradition, proximity to markets, staff availability, quality of housing, complementary businesses, availability of parking and individual preferences by directors. A similar pattern of choice may exist amongst companies in the same business, the prime example being the City of London, which is the centre for banking, insurance and finance. In the 1980s service companies began to choose different locations for different activities within their companies. For example, their headquarter offices might be in central London and the back-office or support facilities might be located on an 'out-of-town' business park. The service industry is currently going through a period of structural change and re-organization following the recession of the early 1990s. Many companies have introduced the concept of 'hot-desking', i.e. several employees share a desk or workstation – particularly with sales staff who spend most of their time out of the office. There has also been an increase in 'teleworking', i.e. employees working from home via computer and telecommunication links with head office. There is much debate about how such changes in working patterns and the advance in information technology will affect the location of offices in the future.

(b) Retail

Retail development takes place within a hierarchy of shopping locations all over the UK. This hierarchy of shopping locations consists of regional centres, district centres, local centres and superstores/retail warehouses. A particular shopping area will be classified within the hierarchy by reference to its general characteristics and the size of its catchment population. The catchment population is typically calculated by reference to the size of the population living within 10–20 minutes driving time zones from the centre of the area.

A regional shopping centre usually has a catchment population in excess of 90 000 people and a district centre usually has a catchment population of less than 90 000 people. Regional centres are usually large towns and cities, e.g. Bristol and Bath. However, there are a few regional shopping centres in out-of-town locations, e.g. Metro-Centre in Gateshead, Meadowhall in Sheffield and Merry Hill in Dudley. District centres are medium sized shopping centres which tend to cater for weekly shopping needs. Local centres are usually located within residential areas, catering for the daily needs of local residents. Retail warehouses and superstores/hypermarkets are located on main roads on the edge of town centres with good accessibility from nearby residential areas.

A retail developer when seeking a site will analyse the catchment area where the proposed scheme is to be located. In relation to regional and district centres the catchment will be defined in terms of the drive times from the centre and analysed in relation to both the population size and its characteristics in terms of social and economic groups. An analysis will also be made of competing shopping centres and the impact of the scheme on those centres. In carrying out this analysis the developer is assessing the trading potential of the proposed scheme within the hierarchy of shopping centres. In other words, the site is being looked at from the point of view of the retail tenant and the 'shopper', as the success of the scheme will depend on this. This type of analysis will be discussed in greater detail in Chapter 7.

In the case of retail warehousing there are many developers who specialize in so-called 'out-of-town' retailing. They develop a good detailed knowledge of the locational requirements of each retailer through their working relationship with them. Many of the retailers who operate in out-of-town stores, particularly the food retailers such as Sainsburys and Tescos, carry out their own developments. The development division of the retail company will identify potential sites and the retail division will consider the potential trading position.

The most important factor in the precise location of any proposed town centre retail scheme, whether it is a single shop unit or a major shopping centre, is the pedestrian flow. Studies can be carried out into the pedestrian flow, which is influenced by car parks, bus and railway stations, pedestrian crossings, and major stores. It is the precise location of a shop or store which determines its rental value and it is crucial the developer gets it right. In town centre locations, shops are either classified as prime, secondary or tertiary by virtue of the pedestrian flow and proximity to the major stores.

Since the mid 1980s retail development activity has been concentrated in 'out-of-town' and 'edge-of-town' locations with schemes such as Meadowhall in Sheffield and Merry Hill in Dudley becoming the most successful trading places for retailers in the country. The majority of retail floorspace is in town and city centre locations. However, many town centres have lost trade to competing out-of-town schemes and the government responded in 1993 by issuing revised planning guidelines for retail development. Planning Policy Guidance Note 6 (PPG6) (see

Chapter 5 for further details) places greater emphasis on town centre and edge-of-town locations for future retail development. As a result of the revised guidelines it is more difficult for developers to obtain planning consent for out-of-town retailing unless they can provide evidence that the nature of the development does not affect the viability of the town centre.

(c) Industrial

In general terms industrial development tends to take place in the south of England and the Midlands, although it does take place in the regions of Wales, Scotland and the North of England, sometimes with the assistance of government incentives. There are various types of industrial property, each of which have different locational characteristics. Industrial property can be categorized into warehousing/distribution, light industrial and general industrial.

A warehouse is a building used for storage or distribution. Warehouses are occupied by retailers, manufacturers and distribution companies and in recent years their requirements have become very specialized due to technological advances. Accordingly, most warehouse development is carried out on a 'design and build' rather than a speculative basis to suit the individual tenant. In general terms warehousing tends to be located within the Midlands and near London as it is central to all national and regional markets in the UK. Sites suitable for warehousing must be located with good access to the motorway network.

The term light industrial defines processes which involve the manufacture of goods without any environmental impact. There are various types of occupier who require light industrial property: at one end of the scale there is the traditional manufacturer, and at the other there are companies in 'high technology' industries who require office and research and development facilities alongside production facilities. Under current planning legislation (see Chapter 5 and Appendix A), office, light industrial and research and development uses are classified as one class of use, and an occupier can change from one use to another without requiring planning permission from the local authority. Accordingly, light industrial buildings will be developed to a specification and in a location to suit the type of potential occupier. For example, 'high tech' companies who combine their production with their research and development activities will prefer the south of England as they employ graduates who prefer to live there.

General industrial and heavy industrial users tend to be located in the West Midlands, the North and the North West, Wales and Scotland, largely due to tradition, and the lower cost of both premises and labour. Foreign manufacturers and relocating British manufacturers are attracted to the above mentioned regions by the existence of grants and other incentives.

There are some general factors which influence the location of industrial premises. Industrial occupiers need to locate in areas close to their markets and supplies of raw material and with good access to major roads. Any access via secondary roads needs to be good and relatively uncongested. Companies who employ a high proportion of office and research and development staff will often have similar locational requirements to those of office occupiers, i.e. attractive environment and the availability of quality housing.

Once a developer has established the areas of geographical search, the next stage is to define more specifically preferred locations within a town or area. This will enable the developer to establish a strategy and brief for the site finding process. It is important when planning a land acquisition strategy to define the size of target sites. The definition of size might be related to acreage or the investment lot size of the potential development scheme.

2.3   INITIATION

Having researched and defined a strategy for site acquisition the next step is for the developer to actively seek and identify potential development sites. This can be achieved in a number of ways. However, before we discuss the various methods of site finding, it is important to realize that theory and practice often differ. A developer may have a well thought out and thoroughly researched land acquisition strategy but actually achieving that strategy will depend on numerous factors, many beyond the control of the developer. This is where the property development process is unique and a lot depends on the opportunities available. A developer's ability to acquire land is dependent on the availability of land at any particular time. The availability of land is dependent on the state of the market, planning policies and physical factors, and any particular case will also depend on the motives of the particular landowner. The various types of landowners and their motives for owning land have been discussed in Chapter 1. The developer, landowner, agent and public sector are the main actors involved in the initiation process. The landowner may take an active or passive role in the process.

As a general rule there is likely to be more land available during a time when land values are rising rapidly. The availability of land will be influenced by the allocation of land within a local planning authority's 'development plan' and the perceived chances of obtaining planning permission in respect of unallocated areas of land or land allocated for other uses. Although land may be available on the market and is allocated within the development plan for the proposed use, it still might not be suitable for development due to physical factors. The lack of necessary infrastructure such as roads and services might make a development scheme not viable. Also, the state of the ground, which might be contaminated or unstable, may be prohibitive to profitable development. The various ways of initiating the site acquisition process are examined below.

2.3.1   Developer's initiatives

There can be no substitute for knocking on people's doors. A developer may employ an in-house team, an agent or a planning consultant to actively find development sites based on the criteria set out in the site acquisition strategy.

Many developers, particularly those who specialize in a certain type of development (e.g. retail warehousing) and the large house-builders, employ 'acquisition surveyors' or 'land buyers' as a members of their own staff. Their job is to exclusively find and acquire sites in accordance with the company's strategy. The developer needs to acquire a good knowledge of the target area and thoroughly research the relevant planning policies. Finding sites by this method may mean a lot of hard work with no results. Searches will be made by car or foot to identify potential sites. This done, the next step is to find out who owns the land. There are a number of ways of achieving this, which include examining the planning register, contacting the Land Registry, asking local agents or literally knocking on the door.

The planning register is a register of all planning applications and permissions in a particular planning authority's area. When a planning application is made, the owner of the land to which the application relates must receive a statutory notice from the applicant. Accordingly, an examination of the register will the reveal the owner of a particular piece of land, provided a planning application has been made. However, the details of the landowner may be out of date. Local authorities also hold a register of publicly owned vacant and underused land which is also available for inspection. Since 1990 the Land Registry, the statutory registry of all legal titles to freehold and leasehold land in England and Wales, has been open to the public. Previously (except in Scotland), the records of the Land Registry were not available for inspection except by a solicitor making an official search on behalf of a client acquiring land. The Land Registry maintains a statutory register of nearly 15 million properties in England and Wales, which is due to be totally computerized by 1998. A developer can apply to the Land Registry to establish the name and address of the owner of a particular property, if it is registered, for a small fee.

A developer may also employ a planning consultant to carry out a strategic study of a particular area which will involve identifying suitable land within the planning context. A strategic study of an area will involve examining the 'development plans', i.e. structure plans/unitary plans and, where they exist, local plans (see Chapter 5), covering that area, a study of the planning register and discussions with the planning officers of the local authority. The study will usually identify sites which have been allocated in the development plan but not yet developed, commenting on their suitability and availability for the proposed use. A report will be made on each site describing its characteristics, planning history and details of the landowner if known. The study will also identify sites that have not been allocated but where there is a good chance of obtaining planning consent by negotiation or on appeal. The best time to carry out this study is when the development plan is in its draft or review stage as there is then a chance to influence the allocation of land by presenting evidence at the public inquiry to include a particular site. Accordingly, it is of vital importance that developers know the timetable of every review and draft publication of the development plans relevant to their area of search. The study should advise the developer which sites should be pursued.

A developer may employ a particular agent or lobby a number of agents to find sites in a particular area. The developer will need to brief the agent as to his requirements in terms of the nature and size of sites. The agent should have a good knowledge of the area and its planning policies, so very often a local agent is employed or approached. The developer will develop contacts with a number of agents and it is important to develop good relationships to ensure the agent stays loyal. If the agent is retained directly by the developer a fee will be payable if the latter is successful in acquiring a site identified by the agent. This is typically 1% of the land price, but it is very often a matter of negotiation dependent on the agent's subsequent involvement in the development, letting and funding of the scheme. The advantage of using an agent is that they become the developer's eyes and ears. Through their knowledge of the area, they know who owns a particular site and its history. They can anticipate whether a particular site may be coming onto the market for example. With occupied buildings they may know when leases will expire and therefore when possible redevelopment opportunities might arise. They will keep in contact with local landowners and therefore usually know when a site might be coming onto the market.

It is advantageous if sites can be identified early as it gives a developer a chance to negotiate directly with the landowner and secure the site before it goes on the market. A developer's ability to acquire sites off the market will depend not only on negotiating abilities but the state of the market. When the market is booming and land values are rising rapidly the landowner will be strongly advised by agents to put the site on the open market. A negotiated deal may not be possible if the landowner is a local authority, as they are publicly accountable and the District Valuer must be satisfied that the right price has been achieved.

Developers may also identify sites in some less obvious ways. Developers may acquire a whole company as a means of securing a site or an entire portfolio of properties which fit in with their acquisition strategy, e.g. some development companies in the past have acquired retail chains as a means of securing 'prime' high street shop units. In these situations the developer will retain the ownership of the property assets and either sell the operating part of the business or wind the business up. Developers also may acquire individual properties or entire portfolios through direct approaches to other developers or property investment companies.

2.3.2   Agent's introductions

Although agents may be retained exclusively by a developer to find sites, they will also take the initiative and introduce opportunities to developers directly. The opportunity may be a site already on the market or a site that is likely to come on to the market shortly. If the introduction to the developer ends in a successful acquisition of the site, then the agent will expect a fee from the developer, unless they are retained and instructed by the landowner. The fee is typically 1% of the land price but it may be a negotiated fee dependent on the agent's subsequent involvement with the scheme through letting and/or funding instructions.

An agent will introduce a site to the developer who is most likely to be successful in acquiring that particular site. Some agents remain loyal to a developer because that particular developer is an established client and they have established a good working relationship. The agent will look at the experience of development companies and their financial status in making their decision about who to introduce a particular site to. The most likely candidates are those developers who are active in the particular market at the time, whether it be, for example, out-of-town retailing or industrial schemes, and have a history of successfully bidding for sites.

A development company, depending on its size and financial status, may receive introductions on a daily basis when market conditions are favourable. It is particularly important to set up a register of sites that have been introduced to the company because it is likely that different agents will introduce the same scheme to different people within the same company. It is important to avoid duplication of agents, otherwise two acquisition fees might be payable.

When introducing a site to a developer, the agent should provide enough details to enable an initial decision to be made by the developer as to whether or not to pursue the opportunity. Ideally, the particulars should include a site plan, location plan, planning details, and details of the asking price and terms. It is the introducing agent's responsibility to assist the developers throughout the acquisition process. The agent should be able to provide advice on the local property market and rental values to assist the developer in the evaluation process. Information on existing and proposed schemes of a similar nature is also important. The agent will often negotiate the land price on behalf of the developer.

This method of site finding is a two way process. The developer must establish and maintain a good relationship and regular contact with local and national agents. It is important to provide those agents with details of site requirements to avoid a situation where site opportunities are continually rejected and the agent gives up and goes to a rival developer. At the same time, agents should provide a good service to their developer clients. The chances will be that letting and funding instructions will flow from the initial introduction. Other property professionals such as solicitors, planning consultants, architects and quantity surveyors may introduce opportunities to developers. There is some truth in the saying 'it is who you know not what you know' when related to property development.

2.3.3   Landowner initiatives

A landowner may take an active role in initiating the development process by a decision to sell their land or enter into partnership with a developer.

An obvious source for identifying development sites for sale is advertisements, whether in the media, on a site board or via direct mail. All the property publications, e.g. *Estates Gazette*, *Estates Times* and *Property Weekly*, carry advertisements each week for sites and development opportunities.

A developer may receive particulars of a site for sale direct from a landowner or their agent. Sites which are advertised in the open market will involve the developer competing for the site. The degree of competition will depend on how the site is offered to the market. There are various methods including informal tender, formal tender, competitions involving one or several short listings and auctions. The method of disposal will depend on market conditions and the motives of the landowner. The developer may be in competition with any number of others or there may be a selective list of bidders. We discuss below the various methods used by landowners to secure purchasers or partners.

(a) Informal tenders and invitations to offer

Informal tenders or invitations to offer involve inviting interested parties to submit their highest and best bids within a certain time scale. This usually involves all parties who have expressed an interest in the site and the invitation may include an indication of the minimum price acceptable. For example, it might state that offers of over £750 000 are invited and indicate what conditions attached to the bid might or might not be acceptable to the landowner.

The important point from the developer's perspective is that the bid made is subject to any necessary conditions. After a bid has been accepted by the landowner, the developer then has the ability to renegotiate the price if there is some justification to do so before contracts are exchanged. There is always a risk that the landowner may not accept a revised price and may go to one of the other parties who made a bid. Developers prefer informal to formal tenders as they allow bids to be made on the developer's own terms. However, the more conditions a developer attaches to a bid the less likely it is that the bid will be acceptable, even if it is the highest received. The landowner as a general rule will accept the highest bid unless the conditions attached to it are unacceptable or the developer's financial standing is questionable. After receiving the bids, the landowner may negotiate with several of the parties before making a decision in an attempt to vary conditions or the level of the bids.

(b) Formal tender

A formal tender binds both parties to the terms and conditions set out in the tender documentation subject only to contract. It involves an invitation to interested parties to submit their highest and best bids by a certain deadline. The invitation will set out the conditions applicable. The document will usually state that the landowner is not bound to accept the highest bid. This method may or may not involve a selection of the interested parties. Developers, as a general rule, do not favour formal tenders as it reduces their flexibility and hence increases their risk. The exception to this would be a situation where all the possible unknowns had been eliminated, e.g. where a detailed acceptable planning consent was in place, a full ground and site survey had taken place and the site was being sold with full vacant possession.

(c) Competitions

Competitions are used by the landowner when financial considerations are not the only criteria for disposal of the site. Thus, competitions are mainly used by local authorities and other public bodies seeking to choose a developer to implement a major scheme. They are also used in a more informal way by other landowners seeking development partners. For example, a landowner may want to obtain planning permission before disposing of the land and, therefore, the developer may be selected on the basis of planning expertise. Alternatively, the landowner may not wish to dispose of the land and will seek a developer to project manage the scheme in return for a profit share.

As the majority of competitions involve local authorities and other public bodies, we shall confine ourselves to discussing public authority competitions. Local authorities and other public bodies will invite competitive bids on a tender basis, whether formal or informal, and the bids will normally be judged on a financial and/or a design basis. As a first step, the authority will usually, advertise their intention to set up a competition and invite expressions of interest. Alternatively, the authority may choose a selection of developers to enter the competition.

If the former method is adopted, then usually developers will be invited initially to express their interest in becoming involved. They will usually be asked to provide details of their relevant experience and track record, financial status (usually a copy of their company report and accounts), the professional team if appointed and any other information which is relevant. For example, a developer may own adjoining land to the competition site or may have been involved with the subject site for some considerable time.

The authority will assess the expressions of interest and compile a shortlist of developers to enter the competition. This may or may not be the final selection process, and bids may be invited from those shortlisted in order to compile a final shortlist. The number of selection processes will depend largely on the numbers of interested parties and the complexity of the competition. If the authority is asking developers to submit both financial and design bids, and the design being asked for is fairly detailed, then the number of developers shortlisted for the final process should be no more than about three. Many competitions involve developers in spending large sums of money at risk and in these circumstances long shortlists are not favoured.

It is important that a development brief is prepared to provide guidelines for the competition. The development brief should state the basis of the competition and on what criteria the eventual developer will be chosen. The development brief will set out the requirements of the authority with regard to such matters as total floor space, pedestrian and vehicular access, car parking provision, landscaping and any facilities which the authority considers desirable in planning terms. The authority may include a sketch layout or outline sketch drawings illustrating the development required, but in the majority of cases it is the developer's responsibility to suggest design solutions. The brief should state how flexible the authority will be in assessing whether the bid meets its requirements. It is very important that a developer finds out whether they will be penalized for not strictly adhering to the brief. As a general rule, developers who follow the guidelines set out in the brief will be looked upon favourably, unless a developer proposes a better solution to that outlined in the brief. It may be that through their ability and expertise a developer may produce a higher financial bid by proposing more square footage than that envisaged in the brief whilst still producing a good and sensitive design. Every competition is different and it pays the developer to study the development brief in depth and look at all possible angles that can be used to advantage.

Developers find competitions the least attractive method of acquiring development sites. Competitive bids involving designs necessitate time and expense on the preparation of drawings and financial bids which will prove abortive if the developer does not win.

(d) Auctions

Some development sites are sold at auction. Many of these are unusual sites. For example, public bodies like British Rail (now Railtrack) use auctions to dispose of disused railway embankments and land with no, or limited, access. Auctions may sell property investment opportunities where the leases are due to expire in the next 5 years, and there is obvious redevelopment potential. It pays a developer regularly to look through auction catalogues for development opportunities.

At auction the highest bid secures the site, providing that the reserve price has been exceeded. The landowner will instruct the auctioneer of the reserve price which is effectively the lowest price acceptable. If the reserve price is not reached through the bidding then that particular lot is withdrawn. The auction will set out the standard conditions of sale and special conditions of sale relating to each particular lot. Once a bid has been accepted, the successful bidder exchanges contracts at that point by handing over the deposit, together with details of their solicitor. Therefore, if a developer acquires a site at auction a thorough evaluation and all other preparatory work needs to be carried out beforehand. It is possible that the lot may be acquired prior to auction by direct negotiation with the landowner.

Competitions and tenders are the most normal methods of disposal chosen by landowners when market conditions are good. However, a developer will generally prefer to obtain a site off the market, avoiding competition at all costs. If a developer enters a number of competitions and tender situations, they could all be lost or all or some could be won. There is no certainty and the developer may become very frustrated, wasting a lot of time and money in the process. Success is based on the developer's ability to judge the right opportunities to pursue and the right level at which to submit a financial bid. However, in many instances it may be a case of luck and being in the right place at the right time. The site acquisition process is very competitive and it must be realized that even the best thought out acquisition strategy may not be achieved in the way, and in the timescale, first envisaged.

2.3.4 Local authority initiatives

Due to central government policy over the last 15 years the role of the public sector in the development process has become more and more limited. The whole emphasis of central government policy is for local authorities to do the minimum necessary to enable the private sector to participate in development. However, local authorities still have an important role to play in initiating the development process. Their main role is that of planner through the planning system. They may also facilitate development by promoting or participating in development opportunities themselves.

Local authorities are restricted by the scope of their legal powers, the availability of finance and the need for public accountability. Some local authorities are more active than others depending on the political party in overall control of the authority and whether they wish to actively encourage development within their area. We shall now examine the various methods by which local authorities influence the availability of land for development.

(a) Planning allocation

The allocation of land within a local planning authority's development plan establishes the frame work for the permitted use of land and hence establishes its potential value for the purposes of development. In formulating planning policies in the development plan a local authority has to balance the demands of developers against the wider long-term interest of the local community. We shall examine the way in which local authorities allocate land through the framework of development plans and how developers can participate in this process in Chapter 5.

As we have already discussed developers will examine the development plan relevant to the areas identified in their search for sites. Accordingly, the local authority in their role as planner can influence the availability of a particular site by allocating a specific use to it in the development plan. However, it must be stressed that allocating a site in a development plan does not make it available for development. The developer and landowner must be able to agree terms and the site must be suitable in physical terms for the proposed use. Even if it is available it may not be developed because the location of the allocated land does not meet the requirements of occupiers. If the developer and or landowner disagree with a particular allocation within a development plan, in preference to their own site, then they can present their case to a planning inspector at the public enquiry into the development plan.

(b) Land assembly and economic development

Local authorities may make land available for development by assembling development sites for disposal, which may involve their statutory compulsory purchase powers. Their ability to take on this enabling role obviously depends on the amount of land under their control and their attitude towards encouraging development. Some authorities in areas of economic decline and high unemployment are active in encouraging private sector investment. For example, the planning department of Bradford City Council works with developers to bring forward sites for development using their land acquisition and development powers to deal with physical constraints on development. In economically prosperous authorities activity may be restricted to involvement in prestigious sites such as town centre shopping schemes. However, positive participation by local authorities in making land available is not just limited to land assembly, whether by agreement or compulsion, but may include site reclamation; the provision of buildings; the provision of infrastructure/services; the relocation of tenants and general promotion of their area as a business location. Any or all of these activities tend to be described by the general term of 'economic development'.

The need for a local authority to become involved in 'economic development' depends on the initiatives taken by the private sector and whether market forces alone meet the expectations of the local authority for the development of their area through the creation of employment opportunities. A survey by Lancaster University and the Association of District Councils (Audit Commission, 1989) into local authorities operating specific industrial development initiatives between 1984 and 1986 found that more than two thirds undertook property based initiatives (provision of sites and premises). In addition it was found that promotional work was undertaken by more than a half.

However, the extent to which local authorities can undertake 'economic development' is restricted by government policy in relation to local authority finance. The Audit Commission (1989) found that the sums of money spent by local authorities on 'economic development' activity represented less than 1% of overall expenditure. Even though Part 3 of the Local Government and Housing Act 1989 provides the legal power to participate in economic development, local authorities are under no duty to provide services to assist economic development. In addition under the provisions of the same Act, local authorities can only use half of their capital receipts (proceeds from the sale of land or buildings) for new capital investment. The remaining half has to used to redeem debts or has to be set aside to meet future capital commitments or used as a substitute for future borrowing. A local authority's ability to raise money through capital receipts is taken account of when the government sets their credit approval limit, i.e. the extent to which they can borrow money. The definition of capital receipt extends to the receipt of rent (e.g. occupational rents and ground rents) and the receipt of reduced rent in lieu of some benefit (which must be fully valued in monetary terms). In addition temporary financing by local authorities such as the acquisition of land pending disposal to a developer will count against their credit approval limit if the period between acquisition and disposal runs over a year (not unusual in the property development process!).

Some local authorities, particularly those in the inner cities or in areas of high unemployment, receive additional funding from the government. We shall examine the various grants and funding available in Section 2.6.2. However, access to government assistance is becoming increasingly competitive and local authorities are being forced to explore more innovative ways to achieve economic development aims. We shall examine partnerships with the private sector in (d) below.

When a local authority does become involved in land assembly it can be of real benefit to a private sector developer but it will almost certainly lengthen the whole development process. A developer may require the assistance of the local authority when a particular site identified by a developer is owned by a number of different landowners – typical in the case of a town centre location. The local authority may have allocated the site for comprehensive redevelopment in the relevant development plan and therefore is willing to work with the developer to achieve their planning aims. In such a situation the developer may experience difficulties in negotiating reasonable land values with the various landowners as they effectively hold the developer to 'ransom' by demanding unrealistic prices, due to the fact that their landholding is vital to the proposed development. Alternatively, a landowner may be unwilling to sell their landholding as their motivation for occupation of the land is their business. In addition, a particular development site might be land-locked with access under the control of a landowner who is seeking a price well above the market value because of their advantageous position. A local authority who assists with land assembly in this way will try to reach agreement by negotiation. If this is unsuccessful then they may make a compulsory purchase order under the terms of the Local Government Planning and Land Act 1980. Local authorities, subject to the authorization of the Secretary of State, have power to acquire:

1. Any land which is in their area and which is suitable for and is required in order to secure the carrying out of development, redevelopment and improvement.
2. Any land which is in their area and which is necessary to acquire to achieve proper planning of an area in which the land is situated.

The whole process of compulsory purchase is often a very lengthy and drawnout process, involving the local authority in agreeing compensation values with all the interests (any interest in the land including weekly tenancies) affected by the compulsory purchase order. The number of interests involved may be in the hundreds in the case of an inner city redevelopment and even in the thousands in the case of a new road or railway line through an urban area. Notice must be served on all interests involved and details of the scheme publicized. Every compulsory purchase order is subject to a public inquiry, at which all interested parties may present evidence in front of an inspector appointed by the Secretary of State. Then the Secretary of State, if satisfied with the case presented by the local authority, will confirm the order, which will provide the local authority with the necessary powers to acquire the land. Every claim for compensation is then agreed between the valuers acting for the local authority and the affected party based on the six rules of compensation set out in Section 5 of the Land Compensation Act 1961. Local authorities will appoint an independent valuer or the District Valuer to act on their behalf. In most cases the compensation payable by the acquiring authority is the value of the land which would be realized if it was sold on the open market by a willing seller (Rule 2). In addition the affected party may claim disturbance compensation under Rule 6 which is based on the cost of obtaining new accommodation, removal expenses, cost of adapting new premises, loss of profit, etc. In many cases the disturbance compensation may be more than the value of the land. In the event of disagreement between the parties the matter is referred to the Lands Tribunal for resolution. Further appeal on a point of law may be made to the Court of Appeal.

It is not possible in this introductory book to explore the details of the compulsory purchase procedure but there is plenty of further reading suggested in the bibliography. However, it is worth pointing out that the whole basis of compensation has been widely criticized for not adequately compensating affected parties for the losses incurred. Many businesses displaced by a compulsory purchase order experience great difficulties in acquiring alternative premises which are affordable. This is particularly the case with small businesses operating out of outdated premises at a very low rent. The compensation is based on the value of their existing premises as they stand and not the value of new alternative premises. In such instances many businesses unable to find suitable premises are forced to accept extinguishment of their business as an alternative to relocation, entitling them to claim the value of their permanent loss of profit under the heading of disturbance compensation. The length of the entire process also causes problems for affected businesses who are effectively in limbo pending settlement of their claim and acquisition of alternative premises. There has been pressure from the Royal Institution of

Chartered Surveyors and the Confederation of British Industry (CBI) for the government to change the basis of compensation on the compulsory purchase of commercial interests. They are calling for a more generous basis of compensation to include an additional compensation allowance to take account of the fact that affected businesses are not willing sellers and the purchase is compulsory (residential occupiers are entitled to an extra 10% known as a home loss payment). It is believed that this will improve the speed of the negotiation process. They point to the experience in many countries in Europe where the basis of compensation is more reasonable resulting in a reduction both of the time taken and the cost to the taxpayer.

Another problem caused by compulsory purchase is that of blight which very often affects the areas allocated for redevelopment, where the authority has indicated an intention to use compulsory purchase powers. An example of this was the government's proposal (now abandoned) for CrossRail in London which caused blight to properties along the proposed route for some considerable time. Owner occupiers can serve a blight notice to force the local authority to acquire their land if they have been unable to sell their interest or achieve a price equating to the value of the land without blight. However, this option is only open to commercial owner occupiers whose property has a rateable value less than £18 000. There are no such restrictions on residential occupiers.

In return for assisting a developer with site assembly the local authority may negotiate the provision of social facilities or participate in the financial rewards of the eventual development scheme. However, it can be known for some local authorities in their role as planners to use the threat of their compulsory purchase powers in a negative manner to achieve some material benefit in the form of 'planning gain' (see Chapter 5) or amendments to planning applications.

When local authorities dispose of land to developers they usually produce a development brief which outlines how they would like to see the site developed (see Section 2.3.3). It is important that the brief is flexible and not too detailed to allow the developer freedom to react to prevailing market conditions. Where compulsory purchase powers are used the land assembly process may have taken several years by which time market conditions may have completely changed. It is important from a developer's point of view that sites are disposed of on as clean a basis as possible. In other words any problems or constraints which exist with regard to the legal title, services, planning and access should be tackled from the start.

(c) Infrastructure

The provision of supporting infrastructure is critical to the site acquisition process and local authorities play an important role in its provision. Infrastructure is a term used to describe all the services which are necessary to support development, i.e. roads, sewers, open space, schools.

As we have already discussed in this chapter (see Section 2.2), the existence or proposed provision of roads is important in assessing locations for property development. Whilst proposals for a new road will generate pressure for development along its route; new development will also create new traffic pressures on the existing road network. As the existence of infrastructure is so critical to the viability of a particular development scheme it directly influences land values. If the necessary infrastructure does not exist to support a development then a developer will take account of the cost of its provision in the evaluation of the land value (see Chapter 3 for further details on the evaluation process).

Local authorities are largely responsible for deciding the level of infrastructure required and securing its provision. In performing this role they have to determine who is responsible for the cost of its provision. Due to government control on spending local authorities often negotiate agreements with developers (known as planning obligations; see Chapter 5) to secure the provision of new infrastructure, if it is needed to support the development, e.g. the provision of a roundabout to link the development scheme with the existing road network or the provision of public open space for the public. The assessment of future infrastructure requirements at a strategic level is the responsibility of the county council or the unitary authority (see Chapter 5 for an explanation of the structure of local government) as part of the process of preparing development plans. In most cases the county council or unitary authority is the highway authority for their area, although they may delegate some responsibility to district councils. The highway authority will be consulted on all planning applications to establish whether a development can be supported by the existing road infrastructure. The Department of Transport (Highways Agency) is responsible for motorways and trunk roads. The privatized water companies are responsible for the provision of sewers and the water supply, the cost of which is agreed directly with a developer where it relates to a particular development.

Many authorities adopt a more active approach to the provision of infrastructure: they recognize that new roads open up land for development. Land is often assembled in conjunction with the new road so that the authority can benefit from enhanced land values by packaging sites for disposal to the private sector. There is currently much debate about the pressure for development caused by new roads, particularly in the prosperous and environmentally sensitive areas of the country, and how the cost of provision should be allocated. At the moment developers may be required to enter into Section 106 or Section 278 agreements (under the provisions contained in Sections 106 and 278 of the Town and Country Planning Act 1990) with the relevant highway authority or the Department of Transport, to secure a financial contribution to pay for improvements to existing roads required to accommodate traffic from a particular development proposal. The government were forced to withdraw a very controversial proposal in 1993. It was proposed that the Department of Transport should be able to obtain contributions from developers for motorway and trunk road improvements in accordance with a formula related to the amount of traffic which each development would have added to the road being improved. However, the government did issue in March 1994 revised planning guidelines relating to traffic, in the form of Planning Policy Guidance Note 13 (PPG13), which does affect the preparation of policies relating to the location of traffic generating developments in future development plans. The government wish to reduce the need to travel by car by influencing the location of development schemes relative to the existing road network and public transport. They wish to encourage development which is accessible via alternative forms of transport such as walking, cycling and public services. Overall the aim is to encourage local authorities to co-ordinate land use policies and transport infrastructure to achieve a reduction in the need to travel (for further details on PPG13 see Chapter 5). In addition the government have just announced (December 1994) a substantial review of all major road building projects in the light of evidence that new roads create more traffic congestion. The full implications of all these measures on the future location of development is still being assessed.

Many local authorities are trying to promote public transportation, e.g. light railway/tram systems, in their areas to ease traffic congestion. The Supertram is already operating in Manchester and Midland Metro is planned for Birmingham. In order to secure the necessary public funding for such transportation systems, government regulations stipulate that private sector contributions have to be secured in advance. Often developers and landowners with sites that will benefit from the system proposed will be approached. However, there is a limit to how much developers can contribute as any contribution will be deducted from the land value the developer will be able to pay. The same applies to contributions made by developers for road improvements as discussed above.

Many authorities are adopting a very positive approach to improving the infrastructure of their town centres, in the face of competition from out-of-town retailing, with schemes to improve the overall environment and parking provision (which may conflict with the provisions of PPG13). Town centre managers are often appointed to propose and implement improvement schemes in partnership with the private sector. Retailers such as Boots and Marks and Spencer have contributed towards the salaries of town centre managers. However, local authority resources are limited as revealed in the Department of the Environment's report in 1994 called *Vital and Viable Town Centres*. The government are considering a proposal to levy town centre occupiers in order to finance improvements. Such a proposal has been rejected by the Royal Institution of Chartered Surveyors and the British Property Federation who favour a contribution from the Uniform Business Rate.

(d) Partnerships with developers

We will examine later in this chapter the various methods adopted by local authorities in disposing of their landholdings (see Section 2.5). Many authorities will retain a legal interest in the development scheme by only granting a long leasehold interest to the developer, receiving a ground rent usually linked to the long-term success of the scheme. Alternatively, they may only retain an interest until the scheme is complete, granting a building licence for a nominal premium to a developer to enter onto the land and complete the scheme. Under this arrangement the authority sells the freehold to the developer on completion and, therefore, shares in any uplift in values. Although the arrangements above are presented to the public as partnerships, they are not true partnerships as the private sector bears the majority of the risk whilst sharing some of the rewards.

Development schemes involving local authorities often involve lengthy and costly competitions to choose a development partner. The size of such schemes involve a significant risk for the private sector, with substantial sums of money being expended before funding for the development is secured. The legal agreement between local authority and developer often takes a long time to negotiate and is usually subject to the securing of finance by the developer. Many developers are forced to withdraw from such schemes due to lack of funding or because the initial evaluation of the scheme has changed significantly due to the elapse of time. There is often a conflict of motivations with the local authority's social objective and the private sector's profit objective. Also the local authority has a number of roles in the development process which may conflict, such as planner and landowner. A local authority is run by democratic processes which are often lengthy and inflexible compared with the quick decision making of the private sector. It often takes a while for developers to gain confidence amongst the elected members of the local authority only for changes to occur in the personnel and the political party in overall control.

Many authorities, as an alternative to the above arrangements, have entered into joint ventures with developers via companies limited by share or guarantee. Also some have formed wholly owned subsidiaries, for example, enterprise boards to carry out economic development initiatives. However, Part 4 of the Local Government and Housing Act 1989 regulates and restricts the power of local authorities to set up and participate in companies. This Act limits the amount of interest a local authority can have in a company without the company being treated for accounting and expenditure purposes as an extension of the local authority. Central government wants to ensure that local authorities remain accountable to the public and stops them using joint companies as a means to avoid capital expenditure restrictions. Developers normally prefer joint venture arrangements as they are familiar to them and allow quicker decision making. In addition, they allow more flexibility in securing funding for the scheme as the developer passes some risk to the local authority who shares in any decrease in value of the scheme.

Birmingham Heartlands (now a UDC; see Section 2.8.1) is an example of a joint venture company formed in 1986, between a local authority – Birmingham City Council – and a consortium of private developers. Initially, the shareholding was 65% private and 35% public, which included a 1% share by the local Chamber of Commerce. The company's aim is to regenerate 2350 acres of land to the north east of Birmingham. Local authority involvement includes the provision of new infrastructure, land assembly (including compulsory purchase powers), the promoting of development opportunities and the negotiation of grant aid. The company had to be designated as a UDC in order to secure the necessary finance from central government.

As we will see later in this chapter, central government has recently relaxed Treasury rules concerning financial arrangements made with the private sector by the public agencies responsible for urban regeneration (UDCs and English Partnerships). There is currently much pressure for such relaxation of the rules to be extended to local authorities.

2.4   SITE INVESTIGATION

We have examined the various methods by which developers identify and secure sites. It is now important to examine the various investigations a developer should make before acquiring a site. These investigations will affect the type of conditions the developer will seek to impose in the contract to acquire the site. Although landowners, particularly local authorities, will provide as much information as possible, it is up to developers to satisfy themselves before entering into a commitment to acquire a particular site. The following investigations are of vital importance when a developer is acquiring a particular site. The investigations may reveal that the proposed scheme is no longer viable due to the physical state of the ground and the cost of remedying the problems.

2.4.1   Site survey

A site survey needs to be undertaken by qualified land surveyors to establish the extent of the site and whether the boundaries agree with those shown in the legal title deed. This is of vital importance where a site is being assembled by bringing together various parcels of land in various ownerships. In this instance the survey needs to establish that all the boundaries of the various parcels dovetail together and that the whole of the site is in fact being acquired. It would be a disaster if the developer discovered half-way through a development scheme that a small but vital part of the site had not been acquired. The developer would then have to negotiate from a very weak position with that landowner effectively being held to 'ransom'. The site survey also establishes the contours and levels of the site. If any buildings exist on the site which are to be retained a structural survey will need to be carried out.

The legal search of the title deeds need to establish who is responsible for the maintenance of the boundaries. In addition, the access arrangements to the site need to be checked to ensure that the site boundary abuts the public highway. If a public highway exists the solicitor needs to check whether it has been adopted by the local authority and is maintainable at their expense. If access to the site is via a private road then the ownership and rights over that road need to be established.

2.4.2   Ground investigation

Unless reliable information already exists as to the state of the ground then a ground investigation needs to be carried out by appropriate specialists.

Ground investigations can vary both in cost and extent depending on the proposed scheme and information already known. An investigation will usually take the form of a whole series of bore holes and trial pits taken at strategic locations on the site. Samples taken from the boreholes and trial pits need to be analysed in a laboratory to establish the nature of the soil, substrata and water table, together with the existence of any contamination. The issue of contamination will be examined further below.

The results of the investigation will be given to the structural engineer, architect and quantity surveyor. They will need to analyse the results to establish whether any remedial work is necessary to improve the ground conditions or whether any piled foundations are required, e.g. where the ground is made up with fill material. Both circumstances will have an impact on the cost of the development scheme, which may affect the overall viability.

2.4.3   Contamination

The existence of any contamination on a site has become an issue that developers cannot ignore due to recent legislation and court cases following directives from the European Union (EU). Contaminated land is defined by the government as 'land which represents a natural or potential hazard to health or to the environment as a result of current or previous uses' (Ball and Bell, 1991).

The Environmental Protection Act 1990 brought in a provision under Section 143 which required local authorities to compile and maintain public registers of past and present 'contaminative' uses of land. After several wide ranging consultations on the type of regulations needed to define such uses, the government were forced to abandon their plans for registers in 1993 due to the outcry from the property industry and the business community. It was argued that the existence of such registers would lead to extensive blight of land registered as potentially contaminated. In law the rule 'caveat emptor' (buyer beware) applies to all purchasers of land and the government was trying to provide a means of warning purchasers of land of possible contamination so that polluters did not pass on their responsibilities. The whole emphasis of government policy on environmental legislation is based on the principle that 'the polluter pays', following the example set by EU legislation. The government's review of their previous proposals has recently (December 1994) resulted in a paper entitled *Framework for Contaminated Land* being issued. This paper confirms that Section 143 of the Environmental Protection

Act 1990 will be repealed but instead guidance will be issued by the government to help both the identification of contamination and decisions on the extent of remedial action. The paper recommends that local authorities should have a duty to inspect their area and record all 'significant pollution', identifying the necessary remedial action.

The result of all this uncertainty during the early 1990s has led to financiers and purchasers of development schemes demanding evidence from developers that sites are not contaminated and, where they are, that satisfactory remedial action has been taken. A developer will not obtain finance for a scheme if there is the slightest risk of contamination.

The cost of ground investigation is much higher than normal when contamination exists, representing a potentially abortive up-front cost for the developer. As much information should be obtained on the site's history of uses before any ground investigation is started. This is achieved by looking at ordnance survey maps, local authority records and any other likely source of information, such as previous owners. In areas of the country where contamination is widespread then the local authority may have already compiled records of contaminated land, e.g. Sandwell Metropolitan Borough Council in the Black Country, where up to 80% of sites are contaminated. Local authorities also hold information on waste disposal sites in their area. However, information obtained from records may be limited and will always need to be checked.

The ground investigation will involve taking soil samples down to the water table level, and extensive surveys of all underground and surrounding surface water due to the risk of contaminants seeping through water. The results of the ground investigation will enable an assessment to be made of the extent and cost of remedial measures. In the UK guidelines require the site to be treated in a manner specific to the proposed use and the surrounding areas, which has been reaffirmed by the government's paper 'Framework for Contaminated Land'. There are many different ways of treating contamination including simply removing it, treating it in-situ or containing it under a blanket of clean earth. On-going measures may be required once the development is complete such as venting methane gases to the surface. If contamination is limited to one area of the site it may be possible to design around the problem, e.g. by locating a car park there. If ground has to be filled with imported material as part of the process then, depending on the standard of treatment required, deep piled foundations may be required.

In general terms when a developer is faced with a contaminated site then the remedial measures, typically costing between £100 000 to £500 000 an acre, rule out all but the higher value uses such as retail warehousing. However, grant assistance is available in certain areas, which we will examine in greater detail in Section 2.6.2.

In view of the increasing concern about contaminated land and the debate about who should be financially responsible, it is worth developers spending time at the outset to thoroughly investigate its existence. The appropriate professionals should undertake a full environmental audit, which can then be presented to potential purchasers, financiers and tenants.

2.4.4   Services

The site survey should establish the existence of services (water, gas, electricity and drainage). All the utility companies should be contacted to establish that the services surveyed correspond with each of the companies' records. In addition, the capacity and capability of the existing services to meet the needs of the proposed development should be ascertained. If the existing services are insufficient then the developer will need to negotiate with the company concerned to establish the cost of upgrading or providing new services, e.g. a new electricity substation. Where work needs to be carried out by either an electricity or gas company the developer will be charged the full cost of the work, but a partial rebate is usually available once the development is occupied and the company is receiving a minimum level of income. The route of a particular service may need to be diverted to allow the proposed development to take place. The cost of diversion and the time it is likely to take to complete should be established at the earliest possible stage. The legal search of the title deeds will reveal if any adjoining occupiers have rights to connect to or enjoy services crossing the site. The developer may need to renegotiate the benefits of these rights if they affect the development scheme.

2.4.5   Legal title

Solicitors will be appointed by the developer to deduce the legal title to be acquired and to carry out all the necessary enquiries and searches before contracts are entered into with the landowner. The developer's solicitor will apply to the Land Registry to examine the official register of the title. If the land to be acquired is leasehold the register will reveal brief particulars of the lease and the date it was entered into. The developer will need to establish the length of the lease, the pattern of the rent reviews and the main provisions of the lease. Such provisions need to be checked to ensure the terms are acceptable to the provider of development finance. The solicitor needs to establish that the land will be acquired with vacant possession and that there are no unknown tenancies, licences or unauthorized occupancies. The fact that a site or building is unoccupied does not necessarily mean that no legal rights of occupancy exist. The register will also reveal the existence of any conditions or restrictions affecting the rights of the landowner to sell the land. In addition, all rights and interests adversely affecting the title will be established, such as restrictive covenants, easements, mortgages and registered leases.

The existence of an easement might fundamentally affect a development scheme. An easement may be positive, e.g. a public or private right-of-way, or negative, e.g. a right of light for the benefit of an adjoining property. If they exist to the detriment of the proposed scheme then the developer may be able to negotiate their removal or modification to allow the scheme to proceed. Rights of light might affect the proposed position of the scheme and the amount of floorspace. If a party wall exists then it will be necessary to agree a schedule of condition with the adjoining property or make a party-wall award of compensation. A chartered surveyor with specialist knowledge on party wall matters may need to be appointed by the developer.

The existence of restrictive covenants may adversely affect the development scheme, e.g. a covenant prohibiting a particular use of a site. Very often it is difficult to establish who has the benefit of a covenant as the covenant might have been entered into some considerable time ago. If the beneficiary can be found then the developer may be able to negotiate the removal of the restriction. If not, then the developer can apply to the Lands Tribunal for its discharge, which is a lengthy process. Alternatively, the developer can take out an insurance policy to protect against the beneficiary enforcing it. The insurance cover will compensate against the loss in value caused by any successful enforcement action.

A solicitor will carry out a search of the local land charges register maintained by the local authority. This will reveal the existence of any planning permissions or whether any building or site is listed as a building of special architectural or historic interest. Enquiries will also be made of the local authority to establish whether the road providing access to the site is adopted and maintained at public expense. The existence of any proposed road improvement schemes might affect the site, e.g. a strip may be protected at the front of the site for road widening purposes.

Enquiries will also be made of the landowner (vendor) and will include standard questions on matters such as boundaries and services. Enquiries will also reveal the existence of any over-riding interests (rights and interests which do not appear on the register of the title) or adverse rights (rights of occupiers of the land). Solicitors may also make additional enquiries of the vendor which are particular to the land being acquired.

The developer must aim to acquire the freehold or leasehold title of the development site free from as many encumbrances as possible by renegotiating or removing the restrictions and easements. Financiers, particularly the financial institutions, prefer to acquire their legal interest with the minimum of restrictions which might adversely affect the value of their investment in the future. The developer has to be able to 'sell' the title to purchasers and tenants as quickly as possible without complications.

2.4.6   Finance

Finally, no prudent developer (unless there are sufficient internal cash resources) would consider entering into a commitment to acquire a site without first having secured the necessary finance or development partner to at least cover the cost of acquisition, including interest on the acquisition cost, while the site is held pending development. The developer should aim to ensure that the financial arrangements are completed to coincide with the acquisition of the site. If no financial arrangements are in place then the developer must be satisfied that either the finance will be secured or that the site can be sold on the open market if no funding is forthcoming. The developer must ensure that all investigations have been carried out thoroughly so that any financier or partner has a full and complete picture of the site. Every area of doubt must be removed if at all possible.

The developer should also obtain specialist advice on the implications of VAT on any site acquisition. The effect of VAT in relation to the evaluation process is described in Chapter 3.

2.5   SITE ACQUISITION

The findings of all the above investigations should be reflected in the site acquisition arrangements. The degree to which developers reduce the risk inherent in the property development process depends to some extent on the type of transaction agreed at the site acquisition stage. The prudent developer will always endeavour to reduce the element of risk to a minimum and the site acquisition arrangements are important in this respect. Ideally, no acquisition will be made until all the relevant detailed information has been obtained and all problems resolved. In practice, however, it is virtually impossible to remove every element of uncertainty. The degree to which the developer can reduce risk to the site acquisition stage is dependent on the landowner's method of disposal, the amount of competition and the tenure. It is possible to pass some of the risk to the landowner, but this will largely depend on the developer's negotiating abilities.

The majority of site acquisitions are on a straightforward freehold basis. The freehold title transfers from the vendor to the developer once contracts have been completed and from that point onwards the developer is entirely at risk. The developer can reduce the risk inherent in the transaction through negotiation of the contract terms. The contract can be conditioned and payments can be phased or delayed. For instance, if there is no planning consent in existence then the developer should negotiate that the contract is subject to a 'satisfactory planning consent' being obtained. The vendor, if such a condition is acceptable, will try to ensure that the term 'satisfactory planning consent' is clearly defined. The developer may obtain a planning consent which does not reflect the optimum value of the site but satisfies the condition in the contract and then, at a later stage, obtain a better planning consent. It is not uncommon for 'top-up' arrangements to be made whereby the vendor benefits from any improvement created by planning consents obtained by the developer. Developers will carefully weigh the degree of uncertainty in relation to planning and it will be a matter of judgement as to whether the risk involved is acceptable. If the vendor is undertaking to sell the site with vacant possession then the contract should be conditional upon this for there could be a delay in the occupants of a building leaving.

Whilst the normal period between signing a contract to purchase a site and the completion is 28 days, the developer may negotiate a delayed completion, e.g. six months. Delays in the development process cost money, so the developer should ensure that any potential problems revealed by investigations are dealt with before contracts are completed, or the time needed to resolve them is reflected in the evaluation and hence the price paid for the land. As an alternative, especially if the planning process is perceived to be long and difficult, the developer will often consider it advantageous to pay for an option to reserve the land. An option involves the developer paying a nominal sum to secure the right at a future date to purchase the freehold. There is usually a 'long stop' date after which the vendor is free to sell the land to anyone if the developer has not taken up the option. The option agreement might specify that after certain conditions have been complied with the developer has to purchase the land. If the developer fails to complete the purchase by the 'long stop' date then the vendor is free to market the site elsewhere. Alternatively, the agreement may allow the developer to call upon the vendor at any time to sell the site after sufficient notice. The developer will aim to fix the value of the site at the time the option agreement is entered into, but in practice this is difficult to achieve. Conversely, at least in a rising market, the vendor will usually try to ensure that the open market value is fixed at the time the developer actually purchases the land.

The developer may only be able to acquire a long leasehold interest in the land at a premium with a nominal ground rent such as a peppercorn. This happens when the landowner is only able to dispose of a leasehold interest, e.g. the Crown Estate, or wishes to retain some control over the development, e.g. local authorities. The developer may be able to take a building lease in the first instance, which will immediately vest a legal estate, probably subject to covenants relating to the satisfactory completion of the development. Alternatively, the transaction might be arranged on the basis of a building agreement and lease which, in the first instance, merely gives to the developer a licence to enter onto the site and construct the building with a commitment by the landowner to grant a lease when the building has been satisfactorily completed.

A similar arrangement is often made by local authorities in relation to freehold transactions when the developer is able to carry out the development under a building agreement and the freehold is transferred on the satisfactory completion of the development. Under this type of transaction the local authority may become an equity partner in the scheme. The value of the scheme is assessed on completion and, therefore, the authority can share in any growth. The developer can use this type of transaction to reduce risk at the outset. The developer may only be required to pay a nominal premium to enter into a building agreement and the consideration to the local authority may be only payable if a profit is made on completion. The consideration paid to the local authority may be in the form of a profit share or it may be the land value on completion. This method of acquisition is advantageous to the developer where the development scheme is large and likely to take a number of years to complete. The risk to the developer is great and the local authority, due to their interest in the implementation of the scheme, may be willing to be flexible. A prudent developer will not enter such a building agreement on a large scheme without making it conditional upon funding.

100 Victoria Street, Bristol – a 34 000 sq ft (3159 sq m) office development by Development Securities plc. The scheme was completed in 1990 and is let to Brann Direct Marketing who have sub-let one floor to Scottish Widows.

It is important that the building agreement is carefully negotiated as otherwise long arguments can take place on completion in relation to the calculation of profit. Typically, the development scheme will take a number of years to complete and as a result it is quite likely that the personnel and the political party in overall control at the local authority will have changed by the time that the project is completed. This might lead to arguments over matters such as the definition of development costs, that adversely affect the authority's profit share. An alternative to endless negotiations about how to calculate the profit is to form a joint company with the authority. With a joint company the profit is clearly shown in the audited accounts of the company and there is no argument as to what are and what are not acceptable development costs.

Sometimes sites are acquired on the basis of a long leasehold interest with an open market ground rent payable, instead of a premium being payable with a nominal ground rent. These leaseholds are usually for between 99 or 125 years. Financial institutions prefer a term of 125 years on the basis that it permits rebuilding halfway through the term. The ground rent can be reviewed in a number of different ways and might be geared to a percentage of the open market rent of the property, a percentage of the rents received less outgoings or rents receivable. A developer should have regard to the preferences of financial institutions. As a rule, financial institutions prefer freeholds to long leaseholds and this will be reflected in the yield at which the institution values the completed investment. The reviews might be to cleared-site value with planning permission on the assumption that a similar term of lease will be granted at the time of review. Some financial institutions prefer the revised rent of the building to be fixed at the time the ground lease is granted. This means that the ground rent might cease to rise or may even fall towards the end of the term. Ground leases may have a user covenant which will limit the use of the site to a particular planning use class.

2.6   GOVERNMENT ASSISTANCE

What happens when land is not made available or development is not initiated by private market forces due to the fact development is not viable? Development is often not viable due to low market rents and lack of occupier demand in a particular area and/or prohibitively high development costs as a result of the physical condition of a particular site. Typically these areas are within the inner cities or regions in economic decline burdened with high unemployment. Very often the infrastructure in such areas is extremely congested or non-existent. There may also be widespread contamination caused by previous use of the land by heavy industries.

To tackle the problem the government have introduced a number of initiatives over the last 15 years. These have been based on the idea of encouraging redevelopment in the inner cities by injecting public sector money into specific areas of the UK in a number of different ways through the Department of the Environment and the Department of Trade and Industry. The Department of the Environment introduced Urban Priority Areas (UPAs) in March 1988 under the government's 'Action for Cities' initiative known as the Urban Programme. There are a total of 57 inner city local authorities designated as UPAs. The Department of Trade and Industry have defined certain areas of the UK as Assisted Areas which have been reviewed from time to time (the last review was August 1993). There are two types of Assisted Area: development and intermediate, the former having greater funding priority from the government's viewpoint. Development and Intermediate Areas benefit from Regional Selective Assistance grants introduced under the Industrial Development Act 1982 to benefit companies locating and expanding in an Assisted Area, which indirectly benefits developers by providing incentives to occupiers. Assisted Area designation is based on the worst levels of unemployment, both long-term and current. Even areas in the South of England, e.g. Hastings in East Sussex, have been classified as Intermediate Areas.

The whole basis of government policy on urban regeneration up until recently has been 'property led', the concept being that property development would attract new companies into the area, which would provide new jobs, which would in turn benefit the wider community. There has been criticism of this approach due to a failure to link the various government initiatives directly with the local economy of the area being targeted for regeneration. The most often cited example is Canary Wharf in London's Docklands. The government have recently launched a whole series of initiatives which introduce a greater element of competition between the areas targeted for regeneration. In addition greater emphasis is being placed on 'partnerships' between the public and private sector, not only to bring about physical renewal through property development, but to bring greater social benefits to the local community. However, this is set against the background of limited financial resources due to the current government's strict control of public spending.

It is important to put the various initiatives we discuss below into the context of recent funding changes by the government. In April 1994 new regional government offices opened bringing together the Departments of Environment, Trade and Industry, and Transport to provide a single point of contact on a regional basis. Each office is run by regional directors responsible for their own budget known as the Single Regeneration Budget. The priority for each budget is directed towards such initiatives (outlined below) as English Partnerships, UDCs, City Challenge and Housing Action Trusts. The remainder of the budget is allocated on a competitive basis to locally based partnerships between the public and private sector. Funding is more likely to be allocated to those schemes which show an integrated approach to regeneration including social issues. Therefore, the government's whole urban regeneration programme is currently based on an environment of competition and partnership between the private and public sector. Under the government's Private Finance Initiative launched in 1992 the Treasury's financial rules have been relaxed to allow UDCs to form joint venture companies with the private sector on a risk sharing basis. This represents a shift in emphasis to encourage more innovative forms of funding urban regeneration involving more risk sharing by the public sector rather than just handing out grants to developers.

Firstly, we will examine all the various government agencies in existence in England, Wales and Scotland set up to implement the government's policy. Secondly, we will look at the various grants and finance directly available to developers and/or local authorities. Thirdly, we will look at the other initiatives designed to encourage occupiers to an area by removing local authority powers and providing tax incentives. And finally, we will look at initiatives aimed at encouraging land availability.

2.6.1   Government agencies

There are various government agencies which exist to implement and administer urban regeneration policies on behalf of the government. Just to make matters a little confusing there are different agencies in England, Scotland and Wales. Their roles differ, but they all take an active, initiating role in the development process by making land available for development and providing financial assistance, and they may even participate directly in development.

(a) Urban Development Corporations

There are currently 13 UDCs in the UK and the first two were set up in 1981 under the provisions of the Local Government Planning and Land

Act 1980. They are similar to their predecessors the New Town Development Corporations and their objective is to regenerate their designated areas. The existing UDCs are London Docklands, Merseyside, Black Country, Sheffield, Tyne and Wear, Bristol, Cardiff Bay, Leeds, Central Manchester, Trafford Park, Teeside, Plymouth Docklands, and Birmingham Heartlands. Birmingham Heartlands is really a hybrid development corporation as it is a joint company comprising a consortium of private sector developers and Birmingham City Council. It was given UDC status by the government to achieve access to public funding under the government's Treasury rules.

The objective of the UDCs is to regenerate their designated areas within a specified timetable. Their aim is to reclaim and service land, refurbish buildings, and provide adequate infrastructure to encourage private sector development. The UDCs are able to offer practical and sometimes financial assistance to developers and owner-occupiers looking to acquire sites in their area. They are also able to assist companies to relocate or expand in their area, with both grants and subsidies including amenity grants, rent and interest subsidies and property capital grants. Developers are able to apply for grant assistance if their proposed scheme is in a UDC area. The UDC becomes the local planning authority for their designated area, although some delegate the administration of planning applications to the local authority. The Board of a UDC comprises members elected by the Secretary of State for the Department of the Environment, the majority of which come from the private sector. All UDCs, even the most recently formed ones, are due to be wound by March 1998. It is not believed that any further UDCs will be set up and their replacement is being seen as City Challenge and English Partnerships (see below).

In 1993 the government relaxed the Treasury rules governing joint ventures between the private and public sector (not including local authorities). As a result UDCs are able to give rental guarantees, as an alternative to grant aid, to private sector (non-residential) developers for a period not exceeding 5 years. In addition they are able to put land into joint venture companies with developers in return for an equity stake in the development scheme. These measures allow UDCs to use land value, rather than cash or grants, to attract developers and hence reduce the risk involved. This approach is being adopted in the redevelopment of the Leyland DAF site in Birmingham Heartlands and the Royal William Dockyard in Plymouth.

It is going to be difficult to assess the success of UDCs until some years after they have been wound up. They were at the forefront of the government's 'property led' policies and they have been widely criticized for ignoring social issues. There has also been criticisms that many infrastructure projects have been provided too late in the regeneration process (e.g. The London Underground extension to the Jubilee Line), due to delays and cutbacks in the provision of the necessary government funds. However, there have been some successes with such developments as Albert Dock in Liverpool, Newcastle Business Park and The Bristol Spine Road. UDCs, in defence of these criticisms, point to their success in bringing about physical renewal and attracting private inward investment.

(b) English Partnerships

English Partnerships is a national urban regeneration agency set up by the government as a new initiative on urban regeneration in November 1993 (under the provisions of the Leasehold Reform, Housing and Urban Area Act 1993), and it became operational in April 1994. Their objective is to reclaim 150 000 acres (60 700 ha) of derelict land in England. English Partnerships has taken over the administration of City Grant and Derelict Land Grant (see below) from the Department of the Environment and controls the former English Industrial Estates Corporation's portfolio of properties. It is based on a regional structure and the aim of its creation is to bring together all the urban regeneration initiatives (except the UDCs) under one body's responsibility. It works outside the area of the UDCs but may take over some of their work when they are wound up.

The aim of the agency is to work in partnership with the private sector, following the relaxing of the government's Treasury rules governing private/public sector partnerships, which now allow rental guarantees and joint ventures. In addition their aim is to work with local authorities and the local community and create employment relevant to the targeted area. It is currently envisaged that English Partnerships will have compulsory purchase powers but they will only be used as a last resort, which has been widely criticized by the property industry. The government have given the agency the task of simplifying the grants system. Indications are that an entirely different approach will be adopted towards the provision of grants which will be tailored to the funding requirements of the specific project. Guidelines published by English Partnerships in December 1994 indicate that various different funding methods will be used including equity participation, short-term funding and rent guarantees. The agency has the power to develop land but the main role is seen as an enabling one.

The initial land reclamation programme will particularly target closed coalfields in the North East, South Yorkshire and the Midlands. In addition English Partnerships propose to reclaim old industrial sites in the North

West, Merseyside, Black Country and the Potteries. Sites which are to be reclaimed will not only provide land for development but provide environmental improvements such as footpaths and open space. Their aim is to concentrate on European Objective 1 and 2 areas (see Section 2.6.2) together with UPAs, Assisted Areas and rural areas in severe economic decline.

The agency benefits from the property portfolio developed by English Estates (a former government agency), which comprises industrial and commercial property. English Estates concentrated their activities in areas where the government wished to promote job creation and economic development, particularly in the Assisted Areas. Their aim was to encourage involvement by the private sector, private investors and local authorities. Examples of such partnerships include Chatham Maritime Park in Kent and Wavertree Business Park in Liverpool. One of its specific initiatives was the Managed Workspace Programme launched in 1988 providing managed commercial and industrial accommodation in inner city areas. The portfolio provides an asset base for English Partnerships, and, like English Estates, property will be disposed of at the earliest opportunity with capital receipts being reinvested in the agency's activities.

(c) Commission of New Towns

The Commission of New Towns (CNT) is a self financed government elected agency responsible for the disposal of surplus land left after the winding up of the New Towns created under the New Towns Act 1946. The total number of New Towns created was 29 spread across England from Crawley in the south to Washington in the North. They were the predecessors to the current UDCs.

The CNT is a land and property agency with the task of selling as much surplus land as is reasonable together with CNT owned property by 1998. They have 18 000 acres (7284 ha) of development land to sell in plots of 0.25–500 acres (0.1–202 ha). Most of the sites have outline planning consent and all are fully serviced. The former development corporations responsible for the New Towns designated most of their landholdings for a specific use under a masterplan and these designations were transferred to the CNT with the landholdings. If a specific site is designated then the CNT can act as planning authority and grant detailed planning consent in certain of the New Towns. If the site is not designated then the CNT can speed up the planning process through their relationship with the local authority.

The method the CNT adopts to dispose of a site depends on the use and the purchaser. A tender procedure is mainly used, but the CNT can be more flexible on sales to owner occupiers or developers with a pre-let. An independent valuer has to verify that the CNT has achieved the best price obtainable in the current market. In certain circumstances the CNT will enter into marketing agreements with developers who have a particular speciality, and they work together to secure a pre-sale or pre-let. The land is sold when an occupier is secured. One of the main priorities of the CNT is to promote inward investment and promote the New Towns to owner occupiers from overseas.

(d) Welsh Development Agency

The Welsh Development Agency (WDA) was formed in 1976 under the Welsh Development Agency Act 1975. The main purpose of the WDA is to further the economic development of Wales due to the decline of the traditional mining and steel industries. Other aims include providing and maintaining employment, improving the environment and promoting Wales internationally as an industrial location. The WDA is controlled by a board appointed by the Secretary of State for Wales. Their role is seen as complementary and supplementary to the activities of all the other public/private sector agencies in Wales. In common with all other government initiatives the WDA aims to encourage private sector participation and to only intervene where there is need for financial assistance.

The WDA can act as developer with the approval of the Secretary of State for Wales and one of its main priorities is the construction of factories for relocating companies. It has successfully attracted foreign companies such as Bosch and Toyota. Other WDA activities include land reclamation [on average 722 acres (292 ha) of land has been reclaimed each year (Simmons, 1995)] and the provision of infrastructure necessary to attract private sector development. Local authorities also receive assistance with environmental improvements.

(e) Land Authority for Wales

The Land Authority for Wales (LAW), originally set up under The Community Land Act 1975, has powers to acquire land needed for development and dispose of it to private sector developers. Their remit is to acquire land and make it available to developers in circumstances where it is considered that the private sector alone would not make it available. In addition it may undertake necessary infrastructure works but not the erection of buildings. Therefore its role is essentially one of land supply, although in practice this is limited due to the lack of finance: it has to be self financing on a year to year basis.

(f) Scottish Enterprise and Local Enterprise Companies

In 1991 the government, under the provisions of the Enterprise and New Towns (Scotland) Act, replaced the Scottish Development Agency (similar to the WDA but operating in Scotland) with Scottish Enterprise in order to combine its economic development role with the functions of the Training Agency. Many of its roles are contracted out to Local Enterprise Companies (LECs), managed by non-executive boards from the private sector, with the objective of targeting very strictly defined areas within Scotland. Scottish Enterprise controls the finances of LECs and they have wider aims than the former Scottish Development Agency combining economic development with job creation and training. A separate organization called Locate in Scotland has the aim of attracting and promoting inward investment in Scotland.

2.6.2 Funding and grants

Direct government financial assistance via funding or grants is available to local authorities and/or developers proposing urban regeneration schemes in specific areas of the country. There are a number of financial grants available to developers and the public sector to develop derelict or rundown inner-city sites and buildings from the government and Europe directed towards UPAs, Assisted Areas and UDCs. We shall now examine the various types of funding and grants available.

(a) City Challenge

City Challenge was launched by the government in 1991 to target funding towards specific rundown inner city areas in Urban Priority Areas. All the 57 local authorities designated as UPAs are required to define a priority area within their authority. Prior to City Challenge every year each of the local authorities in the Urban Programme made bids to central government for direct funding for urban regeneration schemes on a project basis.

City Challenge has changed the whole basis on which UPA designated local authorities obtain funding for urban regeneration projects. Funding is now directed towards partnerships between the public and private sector with proposals for projects in UPAs on a competitive basis. Under City Challenge public and private sector teams are invited to bid for 5 year grants of £37.5 million each. There have been two rounds of competition which have resulted in 31 City Challenge partnerships each with 5 year programmes. The first 11 'pacemaker' programmes were started in 1992 and the remaining 20 second round programmes started in 1993. Every partnership receives £7.5 million each year for 5 years. This flat rate has been criticized as being too inflexible, with many partnerships facing heavy initial costs and others requiring more money in later stages of their programme. The aim is to attract further private sector investment on top of the grants. Programmes are not just based on physical regeneration through property development but also on social and community provision in line with the government's change in policy. Partnerships are encouraged to work closely with the local community. Example of City Challenge programmes include Derby City Council's 'Derby Pride' and Newcastle upon Tyne's 'West End Partnership'.

(b) City Grant and Derelict Land Reclamation Grant

The grant system in England is administered by English Partnerships (previously it was the responsibility of the Department of the Environment and UDCs), who are currently reviewing these two grants with the aim of simplifying the system. The Welsh Office, Scottish Enterprise and the Department of the Environment in Northern Ireland are responsible for administering the grant system in their respective countries.

The City Grant in England (available until April 1995) together with the Urban Investment Grant in Wales and the Local Enterprise Grant (Urban Project Leg-Up Grant) in Scotland are broadly similar grant schemes available directly to private sector developers. In England the grant is available as a priority in the UPAs. In Wales and Scotland the grant is available as a priority in designated district authority areas, for example Cardiff and Dundee. Grant aid is also available to City Challenge schemes and projects in UDC areas. Other inner city areas also qualify for assistance depending on the particular circumstances of the proposed scheme.

To qualify for a City Grant (or its equivalent in Wales and Scotland), it is necessary for the developer to prove that a development project would not proceed without such a grant and the project will benefit a run down city area. Development projects must be worth over £500 000 in England and £150 000 in Wales (no formal limits apply in Scotland and Northern Ireland) when completed to qualify for grant aid. In addition, funds for the project should come from the private sector. The developer seeking grant assistance has to prove the scheme will provide new jobs, private housing or other benefits. The amount of grant obtainable depends on the gap between the cost (including a reasonable allowance for the developer's profit) and value. Cost must exceed value in order to qualify. The scheme must have outline planning consent before any grant application can be made.

The City Grant is paid direct to the developer and an application must be made to English Partnerships (or relevant government department outside England), except in a UDC area where the application is processed by the corporation itself. Once an application is received, an initial appraisal is made, usually within 4 weeks and the developer will know whether the project is eligible in principle. Usually 10 weeks later the decision is made. It is advisable that developers discuss the scheme with English Partnerships or the UDC to assess its chances before submitting an application and committing time and effort to the application. The developer's appraisal will be checked by quantity surveyors and valuers. The grant appraiser will assess the ratio of the public and private sector investment in the scheme known as the 'leverage'. The aim of the government is to achieve an average ratio of 1:4. It seems that English Partnerships are looking to introduce more flexibility into the way the grant is administered to suit the requirements of individual projects.

Once a grant has been approved and accepted, a legal agreement has to be signed by both parties. There may be a number of conditions in the legal agreement relating to the repayment of the grant in certain circumstances. Clawback conditions tend to be applied where the profit achieved is much higher than expected and does not relate to the risk the developer has experienced. Clawback may occur when the scheme is sold or after 5 years, whichever occurs first. Clawback may be applied in proportion to the leverage or on a 50/50 basis depending on the circumstances. Once work has started on the scheme, the developer has to provide quarterly returns of information and claims with supporting documentation. It is important that the developers realize that when the grant is appraised any difference in land price paid by the developer over open market value (assuming City Grants do not exist) will be disregarded. It is a once-only grant and cannot be increased. Projects where work on site has commenced do not qualify for grant aid as it is assumed they are commercially viable.

A Derelict Land Grant (DLG) is available to both local authorities and private sector companies towards the cost of reclaiming land which has been previously developed and is incapable of beneficial use without treatment. It may also be available in limited circumstances to assist with the cost of reclaiming land which is in a condition detrimental to the environment. To qualify for the grant land must be derelict, i.e. the site must have previously been used for a purpose which has ceased, and it cannot be redeveloped without first reclaiming it. When reclaimed the land should be capable of being developed or otherwise have some amenity value for the community, e.g. nature and historic conservation. It is available in all areas of England and Wales, although the amount of grant varies depending on the location and the applicant. In Assisted Areas the grant is payable to local authorities, the WDA and Scottish LECs at the rate of 100% of eligible costs; all other applicants receive a grant of 80%. If the application is outside any designated area then the grant rate is only 50% irrespective of the applicant. Eligible costs include all those directly associated with the reclamation, e.g. reclamation work, surveys, professional fees, legal expenses and auditors' fees.

(c) Infrastructure grants

Infrastructure grants are available in Assisted Areas in the UK. The grant is discretionary and is given towards the cost of providing or improving basic services to stimulate industrial development. Eligible services include drainage/water supply, roads, gas and electricity. Applications may be made by either the public or private sector for the purpose of industrial development, and must demonstrate the creation of new jobs. The maximum grant level is 30% of the infrastructure costs and is usually paid on completion of the scheme. The scheme is administered by the relevant government department in each part of the UK.

(d) European grants

There are many European grants available from the EU, through such structural funds as the European Regional Development Fund (ERDF), for major industrial development and infrastructure projects. The aim of the ERDF is to reduce regional differences in economic development across the EU. All of the European Union's structural funds are administered in accordance with five priority objectives known as Objectives 1 to 5. ERDF is available under Objectives 1, 2 and 5(a).

The Department of Trade and Industry is responsible for securing ERDF funding which is partly used to fund the government's Regional Selective Assistance to industry with the remainder allocated to specific development programmes proposed by local authorities, UDCs, educational establishments and utility companies. Northern Ireland, Merseyside, and the Highland and Islands in Scotland will receive the highest level of ERDF funds available as Objective 1 locations (underdeveloped regions due to

poor economic performance or high unemployment) until 1999. Areas
with Objective 2 status (regions experiencing structural change with above
average unemployment rates) include many of the industrial regions in the
North, Wales and Scotland and more recently some London Boroughs.
Objective 5(a) areas are underdeveloped rural areas. ERDF money spent
has to be matched pound for pound from other sources, i.e. the fund will
only supply up to 50% of the cost of a project, and must not be used to
replace government funding.

2.6.3   Enterprise Zones

The government have since 1981 designated areas of the UK as Enterprise
Zones in Britain to encourage private sector industrial and commercial
development through the removal of administrative controls and the provision of tax incentives. By 1990, 32 Enterprise zones had been designated,
the first being the Lower Swansea Valley in 1981. Enterprise zones were
first introduced as an experiment and despite the government's announcement in 1987 that there would be no general extension to the designation
of them, several zones have been designated in exceptional circumstances,
e.g. the four zones created in Nottinghamshire and Derbyshire following
the closure of the coal mines. The Enterprise Zone status lasts for 10 years
in each designated area and most of the original zones have now lost their
designation.

The following benefits (available for 10 years from the date of designation) in an Enterprise Zone are of relevance to property developers:

1. Exemption from rates for an occupier of industrial and commercial
   property.
2. Full 100% allowances for corporation and income tax purposes for
   capital expenditure on industrial and commercial buildings.
3. A simplified planning system exists, where schemes conforming to the
   published scheme in the Enterprise Zone do not require planning permission.
4. Other statutory consents required are dealt with quickly.

A scheme in each Enterprise Zone sets out the types of development for
which planning permission is deemed to be granted. A limited number of
matters are reserved for approval by the Enterprise Zone's local authority.
Any development proposal which does not conform with the conditions of
a particular scheme will require planning permission, although the application will be processed quickly.

Apart from the simplified planning system and exemption from rates for occupiers, the main advantage to developers in Enterprise Zones is the tax incentives available for new commercial buildings. When obtaining funding for a particular scheme, the developer is able to offer a tax incentive to the investor or owner-occupier. Enterprise Zone allowances allow 100% tax allowances on investments by individuals and companies with relief at the investor's highest rate of tax. They also allow disposal of the investment after 7 years with no clawback of the tax allowances. Investments can be made directly or through approved Enterprise Zone Trusts, which are authorized unit trusts. Rental income is taxable in the normal way. Only the costs of construction (including professional fees) are tax allowable and not the land cost and other expenses.

There have been many criticisms of Enterprise Zones mainly focusing on their displacement effects, e.g. local occupiers moving into these zones from the surrounding area due to the existence of benefits without creating new jobs. The government have addressed this in the creation of the most recent zones by allowing local authorities to put prohibition clauses into contracts with landowners preventing local companies from doing this. The Merry Hill Shopping Centre in Dudley, in the West Midlands, is an example of a very successful development in an Enterprise Zone, but shops and offices in nearby Dudley Town Centre have suffered as a result.

2.6.4 Simplified Planning Zones

Simplified Planning Zones (SPZs), as the name suggests, are designed to speed up the rate of development in areas where it is needed. Local authorities can use SPZs to give advance planning permission for specified types of development in clearly defined areas. They are similar to the planning system in Enterprise Zones. The advantage to developers is that they know in advance what schemes will be permitted and that they do not need to make a planning application or pay planning fees. The first local authority to introduce an SPZ was Derby. Like Enterprise Zones, their status lasts for only 10 years. Any planning authority can propose an SPZ and the only requirement before formal adoption is to publish a draft scheme for public comment. If objections are raised there may be a public inquiry. A developer can propose an SPZ, and if the authority rejects the proposal the matter can be referred to the Secretary of State for the Environment.

2.6.5   Land availability

The government have always been concerned about the amount of vacant and derelict land within urban areas [according to an English Partnerships' Press Release issued in 1994 the government's 1988 Derelict Land Survey found that 100 000 acres (40 468 ha) of land was derelict], and to date they have introduced two initiatives as follows.

(a) Land registers

In 1980 under the Local Government, Planning and Land Act the government introduced land registers. Local authorities are required to provide the Department of the Environment with a register containing details of unused and underused land owned by public authorities. The register is available for inspection at the Department of the Environment's regional offices for potential purchasers and developers. It contains a description of each site with information on ownership, planning history and constraints to development. The Department of the Environment can assist developers to obtain the release of particular plots of land on the open market. However, the presumption of the government in setting up the register was that ownership was the greatest constraint. In practice developers will approach landowners if a viable development is possible. The greatest constraint preventing many areas of vacant land being developed is lack of occupier demand and the physical condition of the site.

(b) Public Request to Order Disposal

In November 1993 the government launched another initiative to bring public vacant land into use, known as 'public request to order disposal' (PROD). The aim in doing so is to unlock the development potential of 80 000 acres (32 374 ha) of publicly owned vacant land. The government is inviting anyone who is aware of vacant land to contact the Department of the Environment, who will then give the owner 42 days in which to make representations on why the site should not be sold. If the Department of the Environment is not satisfied then it will order the land to be sold by auction, tender or private treaty. It is thought that English Partnerships might use this procedure to avoid compulsory purchase orders. However, most active local authorities have been disposing of surplus land to supplement their finances anyway.

Executive Summary

This chapter has shown that even if a developer has a well researched land acquisition strategy its achievement, in the way and time originally envisaged, is very often beyond the control of the developer. The following pre-conditions need to be in place for the development process to be initiated through land acquisition:

1. The landowner's willingness to sell the land on terms and at a price to enable a viable development to proceed.
2. Planning permission for the proposed development or allocation of the proposed use within the relevant development plan.
3. The existence of infrastructure and services to support the proposed development.
4. The existence (if necessary after appropriate treatment at a reasonable cost) of suitable ground conditions to support the development.
5. The necessary development finance.
6. An known end-user or occupier demand for the proposed development.

If any one of the above pre-conditions are not in place then development will not proceed or subsequent development will represent a considerable risk to the developer. Local authority involvement and government assistance may be available in relation to compulsory acquisition, provision of infrastructure, site reclamation, finance or occupier/investor incentives depending on the nature of the proposed development and its location. Most importantly the requirements of occupiers must be at the forefront of any developer's land acquisition strategy.

REFERENCES

The Audit Commission for Local Authorities in England and Wales (1989) *Urban Regeneration and Economic Development – The Local Government Dimension*, The Audit Commission for Local Authorities in England and Wales.

Ball, S. and Bell, S. (1991) *Environmental Law*, Blackstone Press, London.

Chesterton Research (July 1993) *City Office Markets: The National Picture*, Chesterton Research, London.

English Partnerships Press Release (27th June 1994).

Simmons, M. (1995) Focus: development aid – development agencies. *Estates Gazette*, **9501**, 64–65.

CHAPTER 2 CASE STUDY – WASHINGTON CENTRE, HALESOWEN ROAD, DUDLEY, WEST MIDLANDS

This case study of an industrial development in Dudley, being carried out by Folkes Properties Ltd, illustrates many aspects of the initiation and acquisition stages in the development process we have examined in this chapter. The site was in a derelict state and contaminated when acquired: it was described in the local development plan as an 'eyesore'. In order to carry out a viable development grant assistance was obtained from the Department of the Environment in the form of a City Grant.

THE DEVELOPER

Folkes Properties Ltd is the investment property division of Folkes Group plc. The other divisions within the company manufacture heavy engineering and building products. The property division was created in 1965 as a result of surplus land being available from within the parent company. Now Folkes Properties plc is one of the West Midlands' leading industrial property developers owning 2 million sq ft of buildings. They retain all the properties they develop as investment properties or for their own use within the group of companies. They manage all their developments and properties in-house with a small team of staff. A good relationship exists with all their tenants and through the feed-back they receive they continue to alter their product to suit market needs.

SITE ACQUISITION

The site, which extends to 11 acres, was originally acquired through company acquisition back in 1953, when the parent company owned three quarters of the many manufacturing companies in occupation of the buildings on site. In 1987 the companies were rationalized and the buildings were vacated leaving a site surplus to company requirements. The site was transferred for administrative purposes by the parent company to Folkes Properties Ltd who then initiated the development.

The site at that stage did not present an attractive development proposition (see Figure CS2.1), but it is typical of this area of the West Midlands known as the Black Country. The developer has considerable experience of dealing with such sites.

The site is bordered to the south by housing and a canal; to the east by the A459 Halesowen Road together with a landscaped area. The land to the north and west is predominantly industrial. The site is 4.5 miles away from Junction 3 of the M5. At the time of acquisition there was a site density of 80%, with 160 000 sq ft (14 864 sq m) of derelict steel portal framed forges

Figure CS2.1 Site before development.

and ancillary buildings and it was known that there was a high level of ground contamination. The site had originally been a gas works and the ground contained asbestos and arsenic. In addition there were cellars and concrete bases down to 30 ft (9.14 m) below ground. At the time of acquisition metal scrap was being stolen from the buildings and fly-tipping was taking place. It was valued in its derelict state at £200 000.

INITIATION

The use of the site for housing was considered and rejected by the developer due to the ground conditions, as site reclamation needs to be carried out to a very high standard. There was no demand for retail warehousing due to the existence of the Merry Hill regional shopping centre in Dudley, 2 miles away. The developer knew from recent experience that there was a limited and decreasing supply of high quality industrial space within the Black Country. Traditional industrialists in the area tend to operate from outdated or low specification premises, accessed from congested roads, sometimes paying as little as £1.50 p.s.f. (£10.76 p.s.m.) in rent. Folkes Properties Ltd had in the past improved the specification of each of their developments from their previous one and, as a result, their experience showed that rental growth was better on the higher quality units. Adverts for industrial pre-lets were placed in local newspapers and the response indicated a clear demand for space. In addition the Black Country Development Corporation were proposing the construction of a dual carriageway through the heart of the Black Country which would displace 300 factories.

Folkes Properties approached the Chief Executive and the Chief Planner at the local authority, Dudley Metropolitan Borough Council (Dudley MBC), who had designated the site as an 'eyesore' in the Netherton Local Plan. They explained their proposal to develop a high quality industrial scheme and asked for a grant to carry out a site investigation and chemical analysis which would cost £30 000. Folkes Properties had earned a good reputation with Dudley MBC for producing high quality industrial buildings, having recently completed 90 000 sq ft of new development, attracting some major public companies from outside the area, so the grant was approved.

The results of the ground investigation confirmed the existence of asbestos, recommending its removal along with all other contaminated earth, surface and underground concrete. It would need to be replaced with imported fill material to a depth of 10 ft (3 m) below finished ground level which would be compacted. Piled foundations would be needed over most of the site to support the buildings. Although raft foundations could have been specified, it would have required a very rigid and time consuming method of compaction. It was estimated that piled foundations would cost £5.57 p.s.f. (£59.95 p.s.m.) more than raft foundations. Quotations were sought for the remedial work and it was revealed that the remedial work to the site would cost between £500 000 and £900 000. Surveys also revealed that there were no services or drainage. In addition, storm water had previously drained into the canal, but British Waterways Board (who are responsible for the canal) had revoked the licence. The developer was forced to provide for a new storm drain into the Halesowen Road, some 820 ft (250 m) across open land: British Waterways Board were insisting on a betterment payment of several hundred thousand pounds to reinstate the licence! It was also established that due to the sloping nature of the site, a pumping station for the foul sewer and a retaining wall would be required.

The cost of the land reclamation work and all the other additional costs were used in a financial evaluation of the scheme. It became evident that the scheme was not viable. Accordingly, a decision was made to make an application to the Department of the Environment for a City Grant. The site qualified for grant assistance as it is located in the borough of Dudley Metropolitan Council, one of the Government's Urban Programme Areas (see Section 2.6).

By the time the formal grant application was made on 13th February 1989 Folkes Properties had approached the owner of the adjacent site, comprising 1. 4 acres (0.56 ha) to the south of the scheme and had reached agreement to acquire the site in April 1989 for £230 000. The decision had been made to acquire this site as it was in a similar derelict state, and would have been a blight on the proposed scheme on the important main road frontage (see Figure CS2.1). Also, an outline planning application had been made to the local authority for a light industrial scheme and the approval was imminent.

Folkes Properties approached the Department of the Environment direct, having previously made two successful applications on other sites. In a meeting at their own offices they outlined their proposal to develop a high specification light industrial scheme, with the ability to take cranage (important in the Black Country industrial market) and with full specification offices. They proposed a two-tier application: firstly to assist with the ground work and secondly to help with the shortfall in rents achievable. Although, Folkes Properties were proposing a scheme to a much higher specification than was usual in the Black Country they were not anticipating any increase above current market rents. In that one meeting the Department of the Environment agreed the format for the application with the developer; they were able to sit down around the computer, see the figures in the financial appraisal and alter them instantly. The developer had produced their own software package to produce the appraisal.

The grant application proposed a development of 208 000 sq ft (19 323 sq m) of light industrial space, with a 20% office content (on the soon to be combined 12.4 acre (5 ha) site), predicting the creation of 1000 new jobs. A grant of £4 368 000 was applied for on the basis of private sector funding of £8 356 000 (to be provided entirely from the developer's own resources as is usual with their own developments) which represented a leverage of 1:9. The appraisal in the application estimated rents of £3 p.s.f. (£32.29 p.s.m.) and an investment yield of 8%, as it was anticipated that major public companies would be attracted to the scheme. This resulted in a estimated development value of £7 617 000.

Building costs were estimated at £28.62 p.s.f. (£308 p.s.m.) (February 1989 prices), excluding the land reclamation costs estimated at £740 000 (an average of the quotations received as the true cost is never known until the ground is excavated) and external works, abnormal foundations, and services at £1 537 000. Total costs, including the developer's profit of 11% on cost (the developer's normal criteria is 17% on cost but the Department of the Environment take account of the lower risk involved with grant assistance) were estimated at £12 724 000. This left a deficit of £5 107 000 once deducted from the estimated development value of £7 617 000. After adding back the finance charges saved by the City Grant amounting to £739 000, this resulted in a city grant requirement of £4 368 000.

The grant application was approved in July 1989 in the sum of £3 500 000. This was apportioned between the ground works in the sum of £750 000 (100% grant) and the remainder was allocated to the building costs. The grant was approved for a 2 year building period with any clawback becoming operable after 5 years (see Section 2.6).

IMPLEMENTATION

Work started on site 7 months after the grant application was made. The land reclamation work cost only £550 000 plus an additional unforeseen £110 000 for some foul drainage under the canal connected with the remedial treatment, which meant the full grant was not claimed. Phase 1 of the scheme commenced in February 1990 which comprised almost half the scheme plus all the necessary infrastructure for the entire site, so that fully serviced plots were available on the remainder of the site for potential pre-lets. Tenants of Phase 1 include two multinational manufacturers: Matthey (UK) Ltd and Turnills (UK) Ltd, together with two local companies. Phase 2 followed immediately with a 25 000 sq ft (2323 sq m) unit pre-let to Pirelli (Dayco PTI Ltd). When Phase 3 was due to start, the economic recession was at its height and therefore Folkes Properties made a decision to delay commencement of these speculative units. As the grant was due to expire in the middle of 1991, they approached the Department of the Environment for a 1 year extension to the original grant. An extension was granted for 1 year and subsequently for a further 2 years. The scheme was completed in October 1994 and was then 90% let.

Rents achieved on the units let to date are about £4.25 p.s.f. (£45.75 p.s.m.) (with some incentives granted), although Unit 7 achieved £4.60 p.s.f. (£49.51 p.s.m.) and Unit 1 achieved £5.16 p.s.f. (£55.54 p.s.m.) (both these being pre-lettings). It is estimated that the rent roll in 4 years' time will be £900 000 per annum. However, this is unlikely to trigger the operation of the clawback as inflation is built into the development costs for the purposes of the calculation.

Figure CS2.2 Completed development of Phase 1.

The final scheme (see Figure CS2.2) is completely different to that originally drawn up by the architect. To date four different applications have been made as a result of changes in design and layout. The planning authority have been positive throughout, except when a change in the personnel involved resulted in onerous conditions concerning working hours and use. However, the condition was quickly revoked. Other problems encountered included delays in the performance of the Midlands Electricity Board and it took 18 months to build the additional drains under the canal.

CONCLUSION

The scheme has been described by Dudley MBC as 'the flagship of the borough'. Folkes Properties Ltd can be proud of this accolade as they have produced a high quality development which can only encourage further investment in the area. Their belief, on the basis of sound analysis and experience, in a higher quality specification attracted multinational companies into the area, creating jobs appropriate to the local population. It provides a good example of urban regeneration, showing what can be achieved if all the participants work together and are prepared to be flexible.

3 ────────────────

DEVELOPMENT
APPRAISAL AND RISK

────────────────

3.1 INTRODUCTION

In this chapter we shall examine the way in which development projects are evaluated, focusing upon development appraisal. Evaluation is a constant process: the developer will not just carry out one appraisal prior to the acquisition of a development site but will re-appraise the profitability of the scheme throughout the development process.

Risk is an inherent part of the property development process and we shall also consider how this is assessed as part of the evaluation process. Market research is part of the appraisal process and we refer to this in Chapter 7. Here, however, we describe the conventional technique of development appraisal before introducing the various cash-flow methods including the discounted cash-flow. Finally, we shall look at sensitivity analysis and ways in which uncertainty can be contained in order to reduce risk.

3.2 FINANCIAL EVALUATION

3.2.1 The conventional technique

Conventional techniques of evaluation are comparatively straightforward using a form of 'residual' valuation. Quite simply, total development costs are deducted from the estimated value of the completed development to establish whether the project produces an adequate rate of return, either in terms of a trading profit or in terms of an investment yield. The main variables are:

1. Land price
2. Building cost
3. Rents/prices
4. Interest rates
5. Investment yields
6. Time

A residual valuation is also used to assess the likely costs of producing a development scheme and by deducting these costs from an estimate of the value of the completed development scheme to arrive at an affordable land price. The developer will include an allowance for the required return in assessing the total development costs.

The way in which the above mentioned variables are brought together in a development appraisal can be shown best by an example.

Let us assume a 2 acre site (0.8 ha) is on the market and the vendors are seeking a price of £3 000 000. The site is in a prime location in a town lying close to London's M25 orbital motorway and the vendor has obtained planning consent for 45 000 sq ft (4181 sq m) of offices. A developer through his market research has established that rents are currently £18 p.s.f. (£193.75 p.s.m.) for quality air-conditioned office space. A bank has agreed to provide short-term finance for the scheme at an interest rate of 2% above the bank's base rate of 6.25%, to be compounded quarterly, i.e. effective annual rate of 8.47%. The developer's quantity surveyor has advised him that building costs are currently £80 p.s.f. (£861.11 p.s.m.). The agents have advised the developer that the completed scheme should achieve a yield of 6.5% when sold to an investor. The developer will carry out the following typical conventional evaluation as shown in Example 3.1.

Example 3.1 Profit evaluation

	£	£
(a) *Net Development Value*		
(i) Estimated Rental Value (ERV)		
Net lettable area 38 250 sq ft (3553 sq m)		
@ £18 p.s.f. (£193.75 p.s.m.)	688 500	
(ii) Capitalized @ 6.5% YP in perpetuity	15.38	
	10 589 130	
(iii) Less purchaser's costs @ 2.75%	283 408	
Net Development Value		10 305 722
(b) *Development costs*		
(i) *Land costs*		
Land price	3 000 000	
Stamp duty @ 1%	30 000	
Agent's acquisition fees @ 1%	30 000	
Legal fees on acquisition @ 0.5%	15 000	
		3 075 000

(ii) *Building costs*
Estimated building cost
Gross area 45 000 sq ft (4181sq m)
@ £80 p.s.f. (£861.11 p.s.m.)   3 600 000   3 600 000

(iii) *Professional fees*
Architect @ 5%   180 000
Structural engineer @ 2%   72 000
Quantity surveyor @ 2%   72 000
M&E engineer @ 1.5%   54 000
Project manager @ 2%   72 000
                                                                450 000

Other costs
(iv) Site investigations say   11 000
(v)   Planning fees say   4 000
(vi) Building regulations   20 000
                                                                 35 000

(vii) *Funding fees*
Bank's legal/professional fees say   40 000
Bank's arrangement fee   80 000
Developer's legal fees say   35 000
                                                                155 000

(viii) *Finance costs*
Interest on land costs (£3.075m) over   694 950
development period and void of 30 months
@ 8.25% compounded quarterly = $(1.0206)^{10}$

Interest on building costs, professional fees, 180 200
other costs and funding fees divided by a half
(£4.24m/2) over building period of 12 months
@ 8.25% compounded quarterly = $(1.0206)^4$

Interest on building costs, professional fees, 360 400
other costs and funding fees (£4.24m) over
void period of 12 months @ 8.25%
compounded quarterly = $(1.0206)^4$
                                                              1 235 550

Letting and sale costs
(ix) Letting agents @ 15% ERV   103 275
(x)   Promotion   75 000
(xi) Developer's sale fees @ 1.5% NDV   154 586
(xii) Other costs (see text)
                                                                332 861

Total development costs		8 883 411
(xiii) *Developer's profit*		
Net Development Value	10 305 722	
Less total development costs	8 883 411	
Developer's profit		1 422 311
Developer's profit as % of total development costs	16.01%	
Yield on development cost	7.75%	

We shall now examine each of the elements of the appraisal in detail.

(a) Net Development Value

There are two important variables in establishing a development scheme's value: rent and investment yield.

(i) Rent   The rent that a tenant is likely to pay to occupy the proposed scheme is usually established in consultation with an agent or a valuer. An estimate of the rent must be as realistic as possible and based on a thorough analysis of the present and future market trends. The best source of information is comparable evidence of recent lettings of similar schemes within the town or area, adjusted to reflect differences in age, quality and specification. In the absence of evidence a thorough market research exercise needs to be undertaken (see Chapter 7). The developer must base the estimate on firm evidence or analysis to establish today's rent, i.e. at the date of the appraisal. On no account should the developer use a forecasted or overestimated rent in this type of appraisal.

Rents are usually analysed by reference to a rate per square foot (or per square metre). In the case of an office building, the net area of the building (the internal usable space excluding circulation space and toilets, etc.) needs to be established and is known as the net lettable area. Discounts will be applied to the rate used to take account of areas such as basements (due to no or restricted natural light) and floors with restricted headroom. With an industrial scheme it is usual for the 'gross internal area' to be measured which includes all the internal space within the external walls. Retail units are an exception, where the rent is usually analysed in relation to 21 ft (7 m) zones measured from the front of the shop. The first zone is known as Zone A, the second Zone B and the remainder Zone C. Retail units are valued in relation to the Zone A rent which is the rate applied per square foot (or per square metre) to the area of Zone A. Then half the Zone A rate is applied to the area of Zone B and a quarter of the Zone A rate is applied to the area of Zone C. This zoning is to reflect the fact the most valuable space is at the front of the shop. As most shops do not conform to a standard size and shape, adjustments will be made to the above described analysis to reflect return frontages and unusual shapes, which will be based on the experience of the valuer. Large retail units such as department stores and variety stores, together with shops in small parades, are usually analysed on an overall rate rather than any Zone A rate. All measurements should be in accordance with the guidance contained in the current Royal Institution of Chartered Surveyors Code of Measuring Practice, which sets out the various definitions of measurement.

(ii) **Investment yield** The investment yield is used to discount the future rental income stream in order to calculate the capital value of the development scheme today. From the investment yield a multiplier is derived that can be applied to the future income stream and this is known as the Year's Purchase (YP) in perpetuity. The YP is the reciprocal of the investment yield.

In effect what is happening is that a snapshot of the current rental income is being taken and it is then assumed that income will remain at that level in perpetuity. The growth in the rental income and the risks associated with it are reflected in the multiplier used. For example, if the future income from the scheme is estimated to be £10 000 per annum and the investment yield is 5% then 5 is divided into 100 to calculate the YP of 20, which is then multiplied by the rental income to produce the capital value of the development – in this case £200 000. Therefore the investment yield (5%) is derived by dividing the rental income into the capital value (0.05) and then multiplying by 100 to express it as a percentage.

The investment yield is obtained by analysis of sales of comparable properties to the development scheme being proposed. The yield is a measure of the property investor's perception of the future rental growth against the risk of future uncertainty. In general terms the faster the rent is expected to grow, the lower the yield an investor is prepared to pay at the outset. However, risk factors also need to be taken into account. Yields tend to change with changes in the patterns of rental growth or investor demand. Shop developments tend to attract the lowest yields while industrial developments tend to attract the highest yields, due to their previous history of rental growth. Investment yields are further discussed in Chapter 4.

(iii) Purchaser's costs Whether the developer intends to hold the property as an investment or sell the property on completion, the development value needs to be expressed as a net development value to allow for purchaser's costs such as stamp duty, agent's fees and legal fees (including VAT).

(b) Development costs

(i) Land costs Land costs include the land price which is either the price already negotiated with the landowner or, as in this example, the price being sought by the landowner.

Site acquisition costs and fees normally comprise stamp duty at 1% of the land price, legal fees involved in acquiring the site and any agent's introduction fee. Legal fees are usually between 0.25 and 0.5% of the land price, depending on the complexity of the deal. Agent's fees are normally agreed at 1–2% of the land price, depending on whether the agent is being retained as the letting or funding agent.

If the project is being forward-funded by a financial institution, the developer may have to allow for double stamp duty on the land cost if the developer purchases the site before completing funding arrangements with the financial institution. Stamp duty will be incurred on the initial purchase of the land by the developer and the subsequent transfer to the financial institution.

(ii) Building costs Building costs are estimated by the developer's quantity surveyor and are usually expressed as an overall rate per square foot (or square metre), which is then multiplied by the gross area of the proposed building. The building costs are estimated at the time of the proposed implementation of the development project. Usually no allowance is made for inflation during the building contract period but some developers may inflate building costs in their appraisals, particularly in periods of rapidly rising building costs.

(iii) Professional fees These are normally calculated as a percentage of the building costs, and include the architect, the quantity surveyor, the structural engineer, the mechanical and electrical engineer, and the project manager, if applicable. They are normally 12–13% of the building cost and are based on the scale charges of each profession, a negotiated percentage or a fixed fee. The percentage agreed with each member of the professional team depends on the nature and scale of the development. Small refurbishment schemes normally attract higher percentages than larger, complex development projects. If a developer will need to appoint other professionals, such as a traffic engineer, a landscape architect or a party wall surveyor, then these need to be included in any evaluation of the project.

(iv) Site investigation fees   These include fees for ground investigation and land surveys.

(v) Planning fees   These are the fees involved in making a planning application and securing consent for the development project. They normally only include the fees paid to the local planning authority which are based on the scale and nature of the scheme. A list of such charges can be obtained from the local planning authority.

In the above example planning consent has already been obtained. However, in a situation when obtaining planning permission may prove difficult a developer has to allow for planning consultant fees and, in the event of an appeal, costs such as fees for solicitors, counsel and expert witnesses. In addition, the extra time period involved will need to be reflected in the interest costs.

(vi) Building regulation fees   These are based on a sliding scale based on the final building cost. Details of such fees are available from the Building Control department of the relevant local authority.

(vii) Funding fees   These are the fees involved in arranging development finance and will depend on the method of finance. Example 3.1 is bank financed, so the developer will need to pay the bank's arrangement fees, solicitor's fee and surveyor's fee. These fees are a matter of negotiation but usually reflect the size of the required loan, and may be anything between 3 and 10% of the value of the loan (see Chapter 4 for details).

If the development is to be forward-funded with a financial institution, then developers will pay the fund's agent (if appointed) and solicitor's fees, as well as their own agent (if appointed) and solicitor. The developer may also have to pay the fund's building surveyor's fees to monitor the construction of the building on behalf of their client.

(viii) Finance costs/interest   Interest costs are a critical element of the appraisal and reflect either the actual cost to the developer of borrowing money or the implied or notional opportunity cost (reflecting the investment foregone, i.e. the capital could be earning money elsewhere). In Example 3.1 the development company will be borrowing money from the bank and, as a condition of the loan, will be providing some capital from its own resources. It is assumed that the interest rate charged by the bank and the opportunity cost of the developer's own money is the same.

In order to calculate the interest costs the developer has to estimate the length of the development period up until the building is either let (and income producing) or sold, depending on whether the developer wishes to retain the scheme or not. The developer must allow sufficient time for all the preparation work needed after the site has been acquired but before the building contract starts. Also there must be a careful estimation of the letting/selling period (void period) which is based on a judgement of market conditions.

In Example 3.1, the following development timetable is assumed:

Site acquisition, preparation and pre-conract	6 months
Building contract	12 months
Letting period	6 months from completion
Investment sale period	6 months from letting
Total development period	30 months

The site acquisition is the first commitment and therefore interest is calculated on all site acquisition costs from the date of acquisition to the letting/sale of the building (in Example 3.1 it is 30 months). Once the building contract is signed then most of all the other costs will be incurred at various different times over the building contract period (in Example 3.1 it is 12 months). Accordingly, a 'rule of thumb' method of calculating the interest is adopted which assumes that costs are incurred evenly over the contract period. Therefore all the costs (except promotion and letting costs) are divided in half and then the interest is calculated on that sum over the whole period (in this case 12 months).

Once the building contract is complete, then the interest will continue to accrue on all the building and other costs spent (except some of the promotion costs and letting/sale fees) until the date when it is assumed that the building will be let/sold. In Example 3.1 it is assumed that the building is let within 6 months of completion and then sold to an investor within 12 months of completion. It is further assumed that 6 months rent-free are granted to the tenant. If it were to be assumed that some rental income would be received before the sale of the investment then such income would be included in the appraisal and offset against the interest calculation. In Example 3.1 interest is calculated by using the

Amount of £1 formula for compound interest (Baum and Mackmin, 1989). In order to calculate compound interest on a quarterly basis the interest rate of 8.25% is divided by 4 to obtain the quarterly rate of 2.06%, which produces a compound interest formula of $(1.0206)^n$, where 'n' represents the number of quarters over which the interest is calculated.

(ix) Letting agent's fees   These fees usually are 15% of the rental value achieved at letting if joint agents are involved. If only one agent is involved, then the fee is reduced to 10% of the rental value achieved at letting. In some circumstances, the developer may negotiate a fee with the agent, on an incentive basis. It is usual for the tenant to pay the developer's legal fees relating to the completion of the lease documentation.

(x) Promotion costs   The developer has to make an assessment of the likely sum of money that needs to be spent on promoting the project in order to let the property, and very often it is this element of the evaluation that is underestimated at this initial stage (see Chapter 8 on Promotion).

(xi) Sale costs   Sale fees may need to be included if the developer intends to sell the building once it is fully let. These will include any agent's fees together with those of the developer's solicitor, representing between 1 and 2% of the Net Development Value.

(xii) Other development costs   The inclusion of other costs within the evaluation will depend on the nature of the development and will be specific to the project (e.g. party wall agreements, planning agreements with the local planning authority and rights of light agreements.)

If the developer considers that there may be a void period between completion of the development and letting the property, then costs such as maintenance and insurance will need to be included. If a lengthy void period is anticipated then additional costs will need to be allowed for such as business rates (usually payable at 50% after 6 months on empty properties once assessed) and maintenance/management costs. In addition, if the scheme has been forward funded by a financial institution then rent may be payable to the fund until a letting is achieved under the terms of the funding agreement (see Chapter 4).

(xiii) Developer's profit   The residual in this appraisal is the developer's profit which is usually expressed as a percentage of the total development costs or the net development value. The profit that a developer will require will depend on the degree of risk involved with the scheme. It is difficult to generalize but developers will usually seek between 15 and 25% of the total cost; the percentage rising with the perceived risk (see Section 3.2.4). The profit may contain an element for contingencies.

If the developer is an investor wishing to retain the development then profit may be assessed by reference to the yield on cost (in Example 3.1 it is 7.75%). The yield on cost is the total development cost (excluding profit) divided into the first year's rental income. The resulting yield should be higher than the yield applied to obtain the net development value (which is comparable to the yields on similar standing investments). The difference between the two yields represents the profit to the investor.

In Example 3.1 the land price used in the evaluation is the asking price and is therefore fixed. However, in most cases the developer has to establish the land price that can be afforded in order to enjoy a fixed target rate of profit. The landowner (vendor) will be seeking the highest and best offer and may not quote an asking price or may seek offers in excess of a certain figure. In Example 3.2, we assume the developer wishes to ensure a rate of profit of 20% on total development costs, so a residual land evaluation is carried out to determine the affordable land price.

Example 3.2 Residual valuation

	£	£
Net Development Value		
Estimated Rental Value (ERV)		
Net lettable area 38 250 sq ft (3553 sq m)		
@ £18 p.s.f. (£193.75 p.s.m.)	688 500	
Capitalized @ 6.5% YP in perpetuity	15.38	
	10 589 130	
Less purchaser's costs @ 2.75%	283 408	
Net Development Value		10 305 722
Development costs		
Building costs		
Estimated building cost		
Gross area 45 000 sq ft (4181sq m)		
@ £80 p.s.f. (£861.11 p.s.m.)	3 600 000	3 600 000

Professional fees
Architect @ 5%   180 000
Structural engineer @ 2%   72 000
Quantity surveyor @ 2%   72 000
M&E engineer @ 1.5%   54 000
Project manager @ 2%   72 000
                                                                          450 000

Other Costs
Site investigations say   11 000
Planning fees say   4 000
Building regulations   20 000
                                                                           35 000

Funding fees
Bank's legal/professional fees say   40 000
Bank's arrangement fee   80 000
Developer's legal fees say   35 000
                                                                          155 000

Finance costs
Interest on building costs, professional fees, other 180 200
costs and funding fees divided by a half (£4.24m/2)
over building period of 12 months @ 8.25%
compounded quarterly $= (1.0206)^4$

Interest on building costs, professional fees, other 360 400
costs and funding fees (£4.24m) over void period
of 12 months @ 8.25% compounded quarterly
$= (1.0206)^4$ 540 600

Letting and sale costs
Letting agents @ 15% ERV   103 275
Promotion   75 000
Developer's sale fees @ 1.5% NDV   154 586
                                                                          332 861

Developer's profit
@ 20% on net total development costs   1 022 692
(£5 113 461) excluding land costs and interest
on land costs

Net Total Development Costs   6 136 153

Residue, i.e. NDV less NTDC   4 169 569

This residue is made up of the following elements:

Land price =	1.0
plus costs of acquisition @ 2.5% =	0.025
	1.025
multiplied by cost of interest of holding land for development period and void (30 months) @ 8.25% compounded quarterly $(1.0206)^{10}$ =	1.226
Total land cost	1.257
multiplied by profit on total land cost @ target rate of 20%	1.2
	1.508

The residual land value, i.e. the price the developer can afford to pay to ensure the target rate of profit, is therefore derived as follows:

Residue	4 169 569
Divided by factor to take account of land price, costs of acquisition, interest and profit as calculated above	1.508
Residual land value	2 764 966
say	2 765 000

This calculation can be checked as follows:

Land price	2 765 000
plus cost of acquisition @ 2.5%	69 125
Total land costs	2 834 125
multiplied by interest for 30 months @ 8.25% compounded quarterly $(1.0206)^{10}$ = 1.226	640 512
Total land cost	3 474 637
plus Net Total Development Cost excluding profit	5 113 461
Total development cost	8 588 098
Net Development Value, as above	10 305 722
Less TDC	8 588 098
Developer's profit	1 717 624
Developer's profit on cost	20%

This result confirms that at a land price of £2 765 000 the target level of profit of 20% can reasonably be expected to be achieved.

3.2.2 Cash-flow method

The conventional method of evaluation, as shown in Examples 3.1 and 3.2, has two basic weaknesses. First, it is inflexible in its handling of the timing of expenditure and revenue. As a result the calculation of interest costs is very inaccurate. Second, by relying on single-figure 'best estimates' it hides the uncertainty that lies behind the calculation.

The first of these problems can be overcome by carrying out a cash-flow appraisal which enables the flow of expenditure and revenue to be spread over the period of the development, accordingly presenting a more realistic and accurate assessment of development costs and income against time. Therefore, the conventional evaluation shown in Example 3.1 is presented as a cash-flow appraisal in Example 3.3.

As this example shows, by enabling the expenditure to be allocated more accurately over time, a better assessment can be made of interest costs. The 'rule-of-thumb' conventional evaluation, described above, had assumed that building costs would be spread evenly over the building period. In practice, building and other development costs are seldom spread evenly over the period. In Example 3.3 some of the development costs are incurred before or at the start of the building contract period, e.g. funding fees and some of the professional fees. It is usual for the majority of professional fees to be incurred during the pre-contract stage and early in the building contract period, as most of the design and costing work is carried out then. In Example 3.3, only 40% of the building cost has been incurred after 6 months of the contract – the half way point. The building costs in fact follow a normal 'S'-curve pattern of expenditure as follows:

Months	1	2	3	4	5	6	7	8	9	10	11	12
% Total Costs	3	10	14	22	31	40	48	60	73	85	93	97

The remaining 3% of the costs represents the standard practice of holding a retention under the building contract usually for a period of 6 months. The retention sum and period may vary.

In practice, the quantity surveyor should be consulted to assess the timing of building costs. Computer programs are available to calculate the 'S'-curve for a particular project and convert that into the expenditure flow. The project manager can assist in assessing the flow of other costs directly related to the building costs. The timing of all other costs should be capable of assessment by the developer based on experience.

Example 3.3 Cash-flow method

Months	1	2	3	4	5	6	7	8	9	10	11	12	13	14	15	16	17	18	19	20	21	22	23	24	25	26	27	28	29	30	Total
Cost (£000s)																															
Land cost	3075																														3075
Building cost							108	252	144	288	324	324	288	432	468	432	288	144													3600
Professional fees				25		25		40	20	20	40	50	50	25	25	25	25	20	20	20											450
Other fees	6	5		4			5		8		5	2																			35
Funding fees						120	35																								155
Letting fees																								103							103
Promotion																				20	20	25		10							75
Sale fees																														155	155
VAT paid	539	1	0	5	0	25	26	51	30	54	65	66	59	80	86	80	55	29	4	7	4	4	0	42	0	0	0	0	0	27	1339
VAT reclaimed				-539	-1	0	-5	0	-25	-26	-51	-30	-54	-65	-66	-59	-80	-86	-80	-55	-29	-4	-7	-4	-4	0	-42	0	0	-27	-1339
(a) Sub-total (Month)	3620	6	0	-505	-1	170	169	343	177	336	382	412	343	472	513	478	288	106	-56	-8	-5	26	-7	280	-4	0	-42	0	0	155	7648
(b) Balance B/F	0	3645	3676	3701	3218	3239	3432	3625	3995	4200	4567	4983	5432	5814	6329	6889	7417	7757	7916	7913	7959	8008	8039	8137	8474	8528	8586	8602	8660	8719	
(c) Total (a+b)	3620	3651	3676	3196	3217	3409	3601	3968	4172	4536	4949	5395	5775	6286	6842	7367	7705	7863	7860	7905	7954	8034	8082	8417	8470	8528	8544	8602	8660	8874	8934
(d) Interest	25	25	25	22	22	23	24	27	28	31	34	37	39	43	47	50	52	53	53	54	54	55	55	57	58	58	58	58	59	60	1286
Balance C/F (c+d)	3645	3676	3701	3218	3239	3433	3625	3995	4200	4567	4983	5432	5814	6329	6889	7417	7757	7916	7913	7959	8008	8039	8137	8474	8528	8586	8602	8660	8719	8934	8934

Total development cost	£8 934 000
Net development value	£10 306 000
Developer's profit	£1 372 000
Developer's profit as % cost	15.36%

The cash-flow method enables the developer to allow for such an irregular pattern of cost, giving a more explicit presentation of the flow of expenditure and a more accurate assessment of the cost of interest. In Example 3.3 the total interest figure is higher than calculated in the conventional evaluation in Example 3.1. However, given a different pattern of expenditure with professional fees and funding fees being incurred later on then the total interest figure may well have been lower than in Example 3.1. It is impossible to generalize, which is why the conventional 'rule of thumb' method is so inaccurate. In the cash-flow example interest is calculated on the outstanding balance (including interest) at the end of each month at the rate of 0.68% per month calculated as $^{12}\sqrt{0.00847}$ in order to equate to the effective annual rate in Example 3.1 of 8.47% per annum.

In the above example, the project is a single office development. The advantages of the cash-flow method are more clearly demonstrated in relation to developments where receipts occur during the development period prior to final completion of the entire scheme, e.g. a development of phased industrial units, a major retail scheme, a residential scheme or a large and complex mixed-use development scheme which will take a number of years to complete fully and which will be developed in phases.

There are other advantages to the cash-flow method. It enables the developer to allow for changes in interest rates easily over the development period or for different sources of finance within the appraisal. The development company may choose to apply a different interest rate to reflect the opportunity cost of their own money in the scheme. In addition, this method disciplines the developer to think hard about the cash-flow of the project. It highlights the need to delay payments as much as possible and bring forward receipts where possible and advantageous. It shows the developer that cash-flow is an important tool in maximizing profitability. A developer will certainly have to produce a cash-flow appraisal to satisfy potential sources of finance. Most developers use both conventional and cash-flow techniques. They will use the cash-flow method to calculate the interest cost and input the resultant figure into a conventional evaluation for presentational purposes. The cash-flow method will be used throughout the development period to constantly evaluate the project as costs are incurred.

It is important for the developer to assess any VAT implications in any development appraisal. Since the Finance Bill 1989 VAT is charged at the current standard rate of 17.5% on construction costs and interests in rights over and licences to occupy land. The legislation only applies to commercial buildings and not to residential buildings or other qualifying buildings, e.g. 'relevant' residential or 'relevant' charitable buildings. A vendor (in the case of selling an interest in land or buildings) or landlord (in the case of leases and licences) has to opt to charge tax (referred to in the legislation as 'waiver of exemption').

VAT has an impact on the cash-flow of a development project. Although, in the majority of cases, developers will be able to fully recover all VAT paid on land transactions and construction costs if they elect to charge VAT on the sale of the completed building or on rents from the letting of the completed building, a cash-flow implication arises as there may be a delay between the payment of VAT and its recovery. In Example 3.3 a 3 month delay is assumed but in fact in this case the delays have no effect on the overall interest figure. However, on larger schemes the delay in repayments will almost certainly impact on the interest calculation. The legislation on VAT is very complex and it is beyond the scope of this book to examine all of the implications in detail. However, it is important to stress that a developer must fully assess VAT and its direct or indirect effect on a particular development project when carrying out an appraisal. In particular, VAT may affect the rent the developer may achieve on completion of the project if the likely tenant is in an exempt business (i.e. they are unable to recover VAT), e.g. banks, building societies or insurance companies.

3.2.3 Discounted cash-flow methods

Example 3.3 calculates interest on a month by month basis, so that at any point in the development programme the developer can establish the outstanding debt at that particular time. Alternatively, two other cash-flow techniques can be used, known as the 'Net Terminal Approach' and 'Discounted Cash-Flow' (DCF) methods. The 'Net Terminal Approach' simply calculates the interest in a different way but produces the same result as the normal cash-flow method as Example 3.4 shows. The interest is calculated on each month's total expenditure until the end of the development period, i.e. when the development is let/sold and the debt is repaid. This method, will overstate the amount of debt outstanding at the end of each month and has no advantage to the developer over the normal cash-flow in Example 3.3.

The DCF method is entirely different as it does not calculate interest on the monthly expenditure. Instead, it discounts all costs and income to present day equivalents to establish the value of the profit today rather than at the end of the development. The discount rate used is the cost of borrowing the money and the formula used to convert costs and values to the present day is the 'Present Value of £1', which is $1/(1 + i)^n$, i.e. the

Example 3.4 Net terminal approach

Months	Cash-flow (£000s)	Interest until completion @ 0.68%	Total (£000s)
1	3620	1.2255	4436
2	6	1.2172	7
3	0	1.2090	0
4	− 505	1.2018	− 607
5	− 1	1.1927	− 1
6	170	1.1846	201
7	169	1.1766	199
8	343	1.1687	401
9	177	1.1608	205
10	336	1.1529	387
11	382	1.1452	437
12	412	1.1374	469
13	343	1.1297	387
14	472	1.1220	530
15	513	1.1145	572
16	478	1.1070	529
17	288	1.0995	317
18	106	1.0921	116
19	− 56	1.0847	− 61
20	− 8	1.0774	− 9
21	− 5	1.0701	− 5
22	26	1.0629	28
23	− 7	1.0557	− 7
24	280	1.0486	294
25	− 4	1.0415	− 4
26	0	1.0340	0
27	− 42	1.0275	− 43
28	0	1.0205	0
29	0	1.0136	0
30	155	1.0068	156

	Total Development Cost	8 934
	Net Development Value	10 306
	Profit	1 372

reciprocal of the amount of £1 used for compound interest (see Baum and Mackmin, 1989).

Using the same figures as those contained in Example 3.3 a DCF is calculated in Example 3.5.

Example 3.5 Discounted cash-flow

Months	Cash-flow (£000s)	Present value @ 0.68%	Net present value (£000s)
1	3 620	0.9932	3595
2	6	0.9864	6
3	0	0.9799	0
4	− 505	0.9732	− 491
5	− 1	0.9671	− 1
6	170	0.9602	163
7	169	0.9537	161
8	343	0.9472	325
9	177	0.9408	167
10	336	0.9345	314
11	382	0.9282	355
12	412	0.9219	380
13	343	0.9157	314
14	472	0.9095	429
15	513	0.9033	463
16	478	0.8973	429
17	288	0.8913	257
18	106	0.8852	94
19	− 56	0.8792	− 49
20	− 8	0.8732	− 7
21	− 5	0.8674	− 4
22	26	0.8615	22
23	− 7	0.8557	− 6
24	280	0.8499	238
25	− 4	0.8442	− 3
26	0	0.8384	0
27	− 42	0.8321	− 35
28	0	0.8271	0
29	0	0.8216	0
30	155	0.8160	126
31	− 10 306	0.8105	− 8353
		Net Present Value (Profit)	− 1111
	Net Present Value with interest @ 0.068% for 31 months, i.e. multiplied by 1.2338 =		1371

The main advantage of this approach to the developer is that it allows a subsequent calculation of the 'internal rate of return' (IRR), which is the measure used by some developers to assess the profitability of a scheme as opposed to a percentage return on cost. This method is more likely to be used by investors who wish to retain the development within their portfolio and wish to analyse the return on their investment. In order to calculate the IRR, the discount rate is varied by trial and error to the rate which will discount all the future costs and income back to a present value of 0. The disadvantages of this method is that the method does not show the outstanding debt at a particular time and the profit is today's value rather than the actual sum that will be received at the end of the development.

3.2.4 Identifying uncertainty and risk

Although the cash-flow method provides a more accurate and explicit form of calculation, it still relies upon a set of fixed variables. That is, the elements that make-up the calculation, such as building cost and rent, are presented as selected 'best estimates' without giving a true impression of the range from which they have been selected. If we look more closely at the basic example of a conventional evaluation set out in Example 3.1 above, we can see that it is based on a considerable number of variable factors, as follows:

1. Land costs
2. Rental value
3. Square footage (or metres) of building
4. Investment yield
5. Building cost
6. Professional fees
7. Time – pre-building contract, building and letting/sale periods
8. Short-term rates of interest
9. Agents' fees
10. Promotion costs
11. Other development costs

However, these can be reduced to the following four main variable factors which will most affect the profitability of a development project, after the land price is fixed:

1. Short-term rates of interest
2. Building cost

3. Rental value
4. Investment yield

Developers normally rely to a large extent upon the professional advice of the development team to estimate the cost of the main variable factors outlined above. The quantity surveyor and project manager advise on building costs and related costs. The agent will advise on rental value and investment yield, and hence the likely development value. However, in the end, developers must form their own judgement about the estimates that are made of the variable factors. They have to assess the likely risk of the main variables changing when deciding on the required level of return in the evaluation process. It is important that the developer uses today's rental values and building costs in any development appraisal. It would not be advisable for a developer to predict rental values, even when building costs in the appraisal are inflated at current inflation rates, as this would expose the developer to more risk. It cannot always be assumed that rises in building costs during a development will be saved by rises in values.

The rental income, the investment yield and the building cost are usually the most sensitive variable factors. To fix the level of rent, the developer may be able to secure a pre-let and to fix the investment yield it may be possible to pre-fund the scheme with an appropriate institutional investor. Either or both of these may be achieved before or during the development project. This then leaves the developer with the task of ensuring the project is built within budget and on time. Much of this will depend upon the quality of project management, but in some cases a fixed price building contract may be secured. We will discuss each of these alternatives below and we shall see that in each of these cases, in reducing or effectively sharing the risk the developer must expect to have to limit the potential reward.

A developer should never underestimate the element of risk. In every development scheme risk should be identified and if possible contained or reduced. We will now examine the various ways in which some of the variables may be fixed in order to contain the developer's exposure to risk. It is important to remember that as the development process proceeds the developer's commitment increases and the possibility of variation decreases.

(a) Land cost

As we have already discussed in Chapter 2 the purchase price of the land is the first main financial commitment. Ideally a site should not be purchased until the appropriate planning permission has been obtained and the detailed building cost established. If this is not possible, the developer should try to negotiate a contract which is subject to the obtaining of a satisfactory planning consent. If the outcome of the planning application is uncertain then it may be possible to negotiate an option to agree to purchase the land by a future date once planning permission has been obtained. Alternatively, a joint venture arrangement might be entered into with the landowner whereby the land value plus any accumulated additional 'notional' interest might be calculated at a future date during the development period.

Once the land is purchased the developer is committed to a particular location which cannot be changed. However, the value of the land and any development scheme built upon it might be affected by external physical factors such as a new road. Depending on market conditions the developer may be able to make a profit by simply selling the land prior to the commencement of the development scheme. Once planning consent has been obtained the value of the scheme is established, although further applications may be made to improve the value of the site. However, planning applications take time and any improvement in value that might be obtained needs to be balanced against the costs of holding the site.

(b) Building cost

The building cost is the second main financial commitment and a number of other costs relate directly to its final sum, e.g. professional fees. Once the building contract is signed the developer is committed to a certain cost which invariably will move upwards. Many cost increases experienced during the development period are due to the developer's variations or late production of information by the professionals responsible for design. These are matters over which the developer must exercise tight control. There are some ways of making the building cost more certain by passing all or some of the risk and design responsibility onto the building contractor, although greater certainty of cost usually means a higher building cost. These are described in greater detail in Chapter 6. Project control is of vital importance in preventing both increase in building cost and time delays. The employment of a project manager may be advisable. The developer and/or project manager must constantly question every aspect of the building contract in order to contain any problems as they arise.

(c) Rental value

It is important to obtain the best possible estimate of rental value through an analysis of the market and in consultation with the letting agents (if they are to be appointed).

The uncertainty of achieving the estimated level of rent can be removed if a pre-letting can be achieved. Some developers might not proceed with a particular development until a pre-letting is achieved to reduce a considerable element of the risk involved. For example, with a business park scheme or a large industrial scheme the developer may provide all the necessary infrastructure and landscaping initially and then build each element on a pre-let basis. In addition, the developer may build one or two speculative units to show potential occupiers the type of building that could be provided and adapted to suit their individual requirements. Developers of major shopping schemes will need to secure the major tenants to the large units (referred to as 'anchor' tenants) at an early stage in order to attract retailers to the smaller 'unit' shops.

The benefit of achieving a pre-letting in reducing risk has to be weighed against achieving a potentially higher profit in a rising market. In the time it takes to complete the development scheme rents might rise. In the case of anchor tenants in a shopping scheme the developer may have to pay what is called a 'reverse premium' to the retailer to secure a pre-letting. The cost of such premium must be accounted for in the development appraisal.

The other advantage of securing a pre-letting is to reduce the overall development timetable as the building will be handed over on completion without the uncertainty of a void period and further interest payments. Also the existence of a pre-letting greatly improves the ability of the developer to obtain funding on the most advantageous terms as the scheme will represent a far less riskier proposal.

(d) Short-term interest rates

Unless the scheme is being financed entirely by the developer funding arrangements need to be in place before any major commitment is made. In obtaining the necessary finance to acquire the land and build the scheme, the developer will be exposed to any fluctuations in short-term interest rates. However, the developer may at a cost either fix or cap the interest rate (see Chapter 4). If the developer achieves a forward-funding of the scheme with a financial institution then the interest rate agreed with them may be fixed.

(e) Investment yield

The same comments apply to investment yield as to the rental value in (c) above. Investment yields are determined by the property investment market, varying according to investor demand and rental growth at a particular time. The uncertainty of the yield changing over the period of the development can be removed if the scheme is pre-sold or pre-funded. If a scheme is pre-sold to an owner occupier then the developer is really performing the role of project manager. With a pre-funding the developer secures both short-term and long-term finance by agreeing to sell the completed and let scheme to the financial institution. Although the developer still bears the risk of securing an acceptable tenant on satisfactory terms and controlling building costs, as with pre-letting the terms negotiated prior to the commencement of the scheme are likely to be less favourable to the developer than can be negotiated at the end of the project. The developer's chances of securing pre-funding improve if a pre-letting is in place.

When a developer is deciding on how much to reduce the element of risk, a balance needs to be struck between profit and certainty. In general terms the greater the certainty the lower the potential profit. The level of risk a developer is prepared to accept will depend largely on their motivation. Occupiers, contractors, financial investors and the public sector involved in the property development process will all be looking to reduce risk to an absolute minimum, whereas development companies may be willing to accept a much greater degree of risk in return for higher rewards. The degree of risk is usually directly related to the complexity and scale of the proposed development. At one extreme a small self-contained office block pre-let to a major plc company represents a very limited degree of risk. At the other extreme a substantial degree of risk is involved in assembling, over a long period of time, a large town centre site suitable for a comprehensive mix of uses including shops, offices and residential. It is usual where a high degree of risk is perceived for developers to seek development partners in order to share both the risks and rewards.

Uncertainty is endemic in the process of appraising development opportunities, and great attention needs to be given to pre-project evaluation to identify and evaluate the balance between risk and reward. The cost of such work, and the time it takes, often leads to greater savings of cost and time later on in the project. However, the time available to the developer at the pre-project evaluation stage is usually limited, especially in a competitive tender situation. In this situation, the developer's judgement and expertise is critical. Establishing the economic viability of the scheme and the particular characteristics of the market-place before being committed to the major financial burdens of land and building costs is most important. Only when the evaluation has been prepared and discussed by the development team can a decision be made as to whether or not it is prudent to purchase or lease a particular site and, if so, on what terms and subject to what conditions.

3.2.5 Sensitivity analysis

One of the critical questions posed by the evaluation process is how does a developer measure the uncertainty involved in a scheme and therefore how much profit is required to balance the resultant risk? Developers are often criticized for not sufficiently analysing risk. This is a valid criticism; on the other hand, developers cannot afford to be too conservative as they may never be successful in securing sites. A careful balance has to be struck which relies entirely on the developer's judgement. We shall now examine a methods of analysis available to assist the developer in answering the question posed above.

In the previous section, once the land price is fixed, the main variables of an evaluation were identified as being short-term rates of interest, building cost, rental value and investment yield. In most cases, the financial outcome of the development is more sensitive to their variability than to the variability of the other factors previously mentioned because they are the highest values/costs in the evaluation. For example, a 10% increase in building costs is likely to have a more significant impact on profitability than a similar increase on promotion costs. The name given to the procedure for testing the effect of variability is 'sensitivity analysis' and the following provides a brief introduction to it.

One or more factors in the evaluation or appraisal can be varied and the effect on viability measured and recorded. The procedure can then be repeated and the different results compared. If, for example, we take the appraisal set out in Example 3.1 above, we can carry out the sensitivity analysis shown in Table 3.1.

This analysis shows that the outcome of the appraisal is most sensitive to changes in investment yield, rent and building cost. If, for the purpose of our example, we now assume that a change in investment yield is unlikely within the timescale of the appraisal, we can concentrate upon the effect of possible variations in rent and building cost. Let us suppose that the range of possibilities that the development team think appropriate is a range of

Table 3.1   Percentage variation in developer's profit

Variable (original value)	Original value −10%	Original value +10%
Land price (£3 075 000)	+26.45%	−26.45%
Interest rate (8.25%)	+9.18%	−9.14%
Building costs (£80 p.s.f./£861.11 p.s.m.)	+32.46%	−32.04%
Rent (£18 p.s.f./£193.75 p.s.m.)	−70.12%	+70.91%
Gross area and net lettable area (45 000/4181 and 38 250 sq ft/ 3553 sq m)	−38.48%	+38.48%
Professional fees (12.5%)	+1.04%	−3.56%
Investment yield (6.5%)	+79.16%	−64.76%
Agent's fees (15%)	+1.01%	−0.72%
Promotion (£75 000)	+1.01%	−0.13%
Developer's sale fees (1.5%)	+1.01%	−1.08%
Funding fees (£150 000)	+1.23%	−1.23%
Other costs (£35 000)	+1%	−0.28%

rents of £15–21 p.s.f. per annum and a range of building costs of £70–100 p.s.f.. Remember that at this stage we are talking about possibilities and not probabilities and, therefore, that the range is likely to be rather wide. The following matrix can now be prepared, showing the level of developer's

Building cost (£ p.s.f.)	Rent (£ p.s.f.)						
	15	16	17	18	19	20	21
70	3.88%	10.62%	17.32%	24.01%	30.68%	37.32%	43.94%
75	0.42%	6.93%	13.43%	19.90%	26.35%	32.78%	39.18%
80	−2.83%	3.49%	9.78%	16.05%	22.30%	28.52%	34.73%
85	−5.87%	0.25%	6.35%	12.43%	18.49%	24.54%	30.56%
90	−8.72%	−2.78%	3.14%	9.04%	14.92%	20.79%	26.64%
95	−11.41%	−5.64%	0.11%	5.84%	11.56%	17.26%	22.94%
100	−13.94%	−8.33%	−2.74%	2.83%	8.39%	13.93%	19.46%

profit expressed as a percentage (i.e. profit as a percentage of total development value).

The total range of possible outcomes can be seen to be a developer's profit of −13.94% to +43.94%.

The next step for the developer is to narrow the focus of attention by concentrating on the most probable outcomes. Let us suppose that as a result of discussion among the development team, the outer limits of the ranges of rent and building cost are excluded as being possible but unlikely. The developer can now concentrate on a narrower range of outcomes given in bold above.

Although this represents a substantial focus of probability, the range of possible outcomes still remains wide: a developer's profit that ranges from +3.14% to +26.35% gives an indication of the real uncertainty that lies behind the appraisal. The developer must now try to weigh up the possible outcomes, assigning either objectively or subjectively some probability to each estimate of rent and building cost. In the end, the original 'best estimate' of £18 p.s.f. per annum rent and £80 p.s.f. building cost may be selected, but the context of possibility and uncertainty in which it lies can now be better understood. On the other hand, an attempt may be made to fix one of the variables in one of the ways we discussed above. Let us assume, for example, that a pre-letting is agreed at £18 p.s.f. per annum. This now narrows the range of likely outcomes to +9.04% to +19.9%. On these figures, a maximum profit of +26.35% has fallen to +19.9%, but as a trade-off the minimum level of profit has risen from +3.14% to +9.04% and the degree of uncertainty has been reduced. It is just this kind of trade-off that is made possible by sensitivity analysis and by the understanding of probabilities, particularly when they are matched to the use of cash-flow appraisals.

This is no more than a brief introduction to the idea of sensitivity analysis. For more detailed reading see Byrne (in press).

In carrying out sensitivity analysis at the initial evaluation stage, the developer is weighing up the balance between risk and reward. The level of uncertainty in the project is therefore a most important factor. Uncertainty can be reduced by fixing any of the four variables discussed above in the ways we have discussed above.

Conventional methods of evaluation do not provide an any indication of the uncertainty which is an inherent part of the development process. Whilst cash-flow methods of appraisal overcome the inaccuracies of the conventional approach they still only represent a 'snapshot' of the viability of the scheme. Sensitivity analysis is a tool in the developer's decision-making process and can provide a measurement of the risk of the development scheme. It forces the developer to be more specific about the assumptions made and estimates made. It assists but does not replace a balanced and informed decision-making process.

However, there is a danger in relying too heavily on the figures produced in the financial evaluation of a scheme. A developer must avoid the danger of using the evaluation process to justify a development project which on the face of it looks good – often referred to as a 'gut' feeling. Although the evaluation must be thorough and based on the best possible information it should be approached from the point of view of what can go wrong. Even if the figures indicate a viable scheme the developer should always research the market for the proposed development in the particular location (see Chapter 7). The recent *Mallinson Report* (Royal Institution of Chartered Surveyors, 1994) on property valuation recommends greater appreciation of local economic conditions by valuers. In addition, it is urging valuers to advise their clients of the uncertainty surrounding any valuation, particularly if it is judged to be abnormal. Furthermore, the Royal Institution of Chartered Surveyors wish to take a lead in identifying common methods of measuring and expressing valuation uncertainty.

Executive Summary

We have examined the conventional method of evaluation used by developers to assess the profitability of a development scheme, or the land price which can be afforded, given a required return, on any particular site. It has been recognized that the conventional technique is a very crude and inaccurate method. The inaccuracy in the calculation of interest costs can be overcome by using any of the cash-flow techniques including the net terminal approach and the DCF. However, all the methods discussed only produce a residual figure, based on best estimates at the date of the evaluation, which hides the true uncertainty of the outcome of the development. Sensitivity analysis and a thorough analysis of underlying market conditions improve the assessment of uncertainty and risk.

REFERENCES

Baum, D. and Mackmin, D. (1989) *The Income Approach to Property Valuation*, 3rd edn, Routledge & Kegan Paul, London.

Byrne, P. *Risk and Uncertainty and Decision-Making in Property Development*, 2nd edn, E & FN Spon, London (in press).

Royal Institution of Chartered Surveyors (1994) *The Mallinson Report. Report of the President's Working Party on Commercial Property Valuations*, Royal Institution of Chartered Surveyors, London.

CHAPTER 3 CASE STUDY – OFFICE REFURBISHMENT APPRAISAL

This case study illustrates how uncertainty is an integral part of the property development process. It shows how an appraisal can change dramatically over the period of a development scheme. In this particular case contracts were exchanged to purchase the development site at the peak of the development boom just before the property market crashed. The developer used conventional evaluation techniques to appraise the scheme initially together with a limited form of sensitivity analysis. We will compare the evaluation carried out before the site was purchased with an appraisal of the completed development as it stands today.

BACKGROUND

In September 1989 an office building, originally built in the late 1960s, with vacant possession was put on the market by the owners. Offers for the freehold were invited by the end of October.

The building occupies a prominent corner position on a major 'A' road, providing direct access to a motorway, in the suburbs of one of Britain's cities. It is surrounded mainly by residential properties with some neighbourhood shops. There is only one other office property, which is of a similar age, in the immediate vicinity. Accordingly, although it occupies a prominent position, it is rather isolated from public transport and the nearest town centre. Over 2 miles further up the 'A' road, at its junction with the motorway, is an established location for large headquarter offices with some major companies as occupiers. All the neighbouring towns are established office locations.

The property comprised ground and four upper floors with a basement, totalling approximately 22 000 sq ft (2044 sq m) of net lettable accommodation. The building was of concrete frame construction with brick elevations and a flat roof. A lift served all floors but there was no air conditioning. It was a typical 1960s office building with small floor plates (up to 4000 sq ft/372 sq m) and contained numerous internal structural columns. In addition the height between the floor slabs was limited, ruling out the inclusion of any raised floors in a refurbishment scheme. Most new offices tend to have suspended ceilings and raised floors to contain all the necessary services. There were approximately 40 car surface spaces on site.

The construction subsidiary of a development company was actively seeking an office site to develop for their own occupation. The above opportunity was introduced to them by a local agent who knew of their requirements. It was in their area of search, being only a few miles away from their own offices at that time. Thus, a decision was made to investigate and appraise the various development options of the property.

INITIAL EVALUATION

The contractor identified three options in relation to the property:

1. Minimal renovation and redecoration of the existing property to enable occupation by the company.
2. Refurbishment of the existing property to a standard attractive to financial institutions.
3. Demolition of the existing building and provision of a new quality office building with basement parking.

Discussions were held with the local authority to establish the extent of redevelopment that would be permitted. Two schemes were presented to a planning officer: one proposing a refurbishment scheme increasing the gross area of the building from 27 100 (2518) to 29 700 sq ft (2759.13 sq m) and providing a two level car park to the rear; and the other proposing a new building with a gross external area of 33 000 sq ft (3066 sq m). The officer indicated informally that both schemes would be supported by the planning officers and would be likely to obtain the approval of the planning committee. With a new scheme the planning authority would seek a car parking ratio of one space per 30 sq m of building. In planning terms the design of a new building would need to have regard to the proximity of the residential areas and reflect the prominence of the site.

The contractor then investigated the local market for offices, informally consulting with two national agents, who were familiar with the property, and the local one, who had introduced the property. Two new air conditioned offices in the area had recently let at £26 (£279.83) and £25 p.s.f. (£269.10 p.s.m.), respectively. The agents estimated the rental value of new quality building to be between £26 (£279.86) and £27.50 p.s.f. (£296.01 p.s.m.). It was believed that a new building would also attract financial institutional interest at a purchase price representing a 6.25% yield. The rental value of a refurbished building was estimated to be £24 p.s.f. (£258.33 p.s.m.) provided that the building was refurbished to a high standard. The agents were also confident that an institutional investor would be interested in a refurbishment, despite the fact the building was not in an established location and the specification would not include raised floors. They were confident that due to its prominent position and provided it was let to a good covenant it would achieve a yield of 7% on a sale. If a basic renovation was carried out the resultant investment would not be attractive to an institutional purchaser and therefore the yield would be only 8–8.5%. No formal demand and supply analysis was undertaken.

Several conventional evaluations of each option were undertaken using the above estimates of rent and yield, together with preliminary estimates of the various alternative building costs provided by the quantity surveyor. The appraisals were carried out to establish the residual land value on the basis of a target profit on cost of 15%. Residual land values ranged between

£1 855 000 and £3 120 000, based on rents between £22 (£236.81) and £26 p.s.f. (£279.86 p.s.m.), and yields between 6.25 and 7.25%. The most likely scenario for the refurbishment scheme was considered to be a rental value of £24 p.s.f. (£258.33 p.s.m.) and a yield of 7%, which produced a residual value of £2 450 000. This ignored the potential savings to the company of occupying half the building and the profit they would receive on the building contract which amounted to £284 000. The resultant appraisals were presented to the main board of the parent company with the recommendation to offer a price of £2 725 000 for the property subject to contract and the structural survey not revealing any major defect in the structure of the building or the use of any deleterious materials. This figure was based on the refurbishment option allowing for the savings and a reduced profit of 13.85% on cost. The recommendation was made on the assumption that the purchase would be complete at the end of January 1990. It was assumed that on completion the contractor would occupy half of the building and let the remaining space. An institutional purchaser would be sought for the completed investment on the basis of a sale and leaseback arrangement (see Chapter 4 for an explanation of this arrangement). The contractor was proposing to use internal resources to finance the scheme in the interim.

ACQUISITION

The directors supported an offer of £2 600 000, on the basis of a 7% yield and £25 p.s.f. (£269.10 p.s.m.) rent. This offer was then submitted to the vendor and accepted. Contracts were exchanged at the end of November 1989, a 10% deposit having been paid, and in January 1990 the purchase was complete. An appraisal prepared in January 1990 is shown in Table CS3.1. A detailed planning application had previously been submitted in December 1989 for a refurbishment of the existing building involving a floor extension at every level and new curtain wall elevations to the front and rear of the property. Planning permission was granted at the end of April 1990.

It had been assumed in the appraisal in Table CS3.1 that construction would commence at the beginning of April. However, Britain's economy was by now in a deep recession and property values had slumped. As a direct result new construction orders had fallen and the contractor was facing intense competition on the limited number of tenders that were being offered to them. As a consequence the contractor had to review its decision to move into new premises and no commitment could be given to start work on the scheme.

A decision was made to transfer the property to the development subsidiary within the parent company on 1st October 1990. By this time the contractor had stripped the interior of the building in preparation for the refurbishment. So it was decided that the best way to proceed, in spite of market conditions, was to proceed with a speculative refurbishment. This was considered to be the best way of recouping the investment made already. There was no market for it as a development site.

An evaluation carried out by the development subsidiary at the time predicted a loss of around £1 million. So the aim now was to minimize the potential loss; keeping building costs within a tight budget; and letting the building as quickly as possible on the best possible terms.

IMPLEMENTATION

Construction started in January 1991, after the development company had made substantial changes in order to achieve cost savings, whilst maintaining the quality of the proposed scheme. The specification included air conditioning and a new lift. However, due to the ceiling height restrictions there was insufficient room for raised flooring, which meant the provision of perimeter and floor trucking.

The refurbishment was completed in September1991 at a final building cost of £1 912 300 compared with the original estimate of £2 650 000. Table CS3.2 shows an appraisal prepared in October 1992, just 1 year after completion. To date the building remains unlet over 3 years since the refurbishment was complete. It was sold by the developer in a portfolio of other properties to a financial institution in 1994 and they are currently seeking a single letting of the building.

CONCLUSION

Although the severity of both the economic recession and the slump in property values in the early 1990s was greater than predicted, a thorough market analysis of the proposed redevelopment back in late 1989 would have raised questions over the viability of the proposal even though at that stage half of the scheme was effectively pre-let. Too much reliance was placed on the estimates contained in the original appraisals. In addition, no attempt was made to carry out a cash-flow appraisal so the appraisals were inaccurate in respect of the calculation of interest costs and the exclusion of some costs. The difference between the two appraisals demonstrates dramatically the uncertainty that underlines any development appraisal. The development scheme was not only affected by the slump in the economy and property market, but also a change in the assumptions justifying the original purchase. No appraisal should be carried out in isolation and a thorough market analysis should always be carried out. In this case it would have shown an oversupply of both new and refurbished buildings in the development pipeline. A set of figures will not tell you that the building is in an unestablished location with a specification that compromises on the requirements of occupiers.

Table CS3.1 Appraisal on purchase (January 1990) assuming occupation of half

Net Development Value	£	£	£
Estimated rental value			
20 700 sq ft (1923 sq m) offices			
@ £25 p.s.f. (£269 p.s.m.)			
		518 000	
Basement		19 000	
		537 000	
YP in perpetuity @ 7.25%		13.79	
Gross Development Value	say	7 400 000	
Less purchaser's costs @ 2.8%		201 557	
Net Development Value		7 198 443	
		say	7 198 400
Development Costs			
Purchase price	2 600 000		
Legal fess @ 0.5%	13 000		
Stamp duty @ 1.0%	26 000		
Agent's fees @ 1.0%	26 000		
		2 665 000	
Estimated building costs	2 650 000		
Professional fees @11%	292 000		
		2 942 000	
Letting fee @ 15% on half rent	40 000		
Promotion	25 000		
		65 000	
Notional interest on land costs			
@16% for 12 months	426 000		
Notional interest on half building			
costs @16% for 10 months	196 000		
		622 000	
Void costs 6 months rent on half building		134 000	
Total development costs			6 428 000
Profit			770 400
Profit as % return on cost	11.99%		

Note: Appraisal assumes occupation of half the building by the company and a sale and leaseback arrangement, followed by a 6 month letting void on the remaining half.

Table CS3.2 Appraisal on October 1992 (no interest calculation included)

Net Development Value	£	£	£
Estimated rental value			
23 000 sq ft (2137 sq m) offices			
@ £15 p.s.f. (£161 p.s.m.)		345 000	
YP in perpetuity @ 10%		10	
Gross Development Value	say	3 450 000	
Less purchaser's costs @ 2.75%		92 336	
Net Development Value		3 357 664	
			say 3 357 700
Development costs			
Purchase price	2 600 000		
Stamp duty and land registry	27 401		
Legal fees	10 118		
Agent's fees @1.0%	26 000		
		2 663 519	
Final building cost	1 912 328		
Professional fees @11%	261 455		
		2 173 783	
Other development costs*	42 630		
Insurance	13 000		
Void costs to June 93	61 000		
Empty rates	42 000		
Advert hoarding income	−27 628		
		131 002	
Developer's legal costs on sale	20 000		
Agent's sale fee @1% NDV	33 577		
		53 577	
Promotion until June 93	55 000		
Letting fees @15% ERV	51 750		
		106 750	
Total estimated development cost			5 128 631
Loss (excluding notional interest)			−1 770 931

* These include site investigations, structural survey, planning and building regulation fees, various consultants, electricity sub-station and extra building costs.

Note: Appraisal assumes letting by June 1993 with 12 months rent free and investment sale by June 1994. Notional interest based on an average 8% amounts to £1.7 million.

4

DEVELOPMENT
FINANCE

4.1 INTRODUCTION

Two forms of finance are required for property development: short-term finance to pay for the costs of production (i.e. the purchase of land, building costs, professional fees and promotion costs), and long-term finance to enable developers to repay their short-term borrowing and either retain the property as an investment or realize their profit. Whether a developer retains the property as an investment or sells it to realize profits depends largely on their motivation and the prevailing market conditions.

In this chapter we will examine the various sources of finance available to developers and the various methods of finance with some worked examples.

4.2 SOURCES OF FINANCE

Traditionally UK clearing banks and merchant banks have been the providers of short-term development finance, with long-term investment finance being provided by the financial institutions (insurance companies and pension funds) and property investment companies. However, sometimes the financial institutions also take on the role of short-term financier by forward-funding development schemes: they provide the necessary interim development finance to a developer and agree to purchase the property on completion of the scheme. It is important to briefly review the history of property financing.

4.2.1 Historical perspective

The role of the various financiers within the property development process has varied depending on the position of both the business and property development cycle at any particular time in relation to the credit cycle (see Section 1.4). It is important to appreciate that financiers are in the business of making money and property is only one of a number of assets they can invest in and lend money against. Also, each of the various financiers mentioned above will have different motivations and liabilities, influencing their policy towards property as either an investment or as a security against a loan. Since the 1960s developers have been able to move from one financial source to another depending on the investor/lender attitude prevailing at any particular time.

Before the 1960s the roles of the short-term financiers (the banks) and the long-term financiers (mainly insurance companies then) were quite distinct. Developers tended to retain their completed developments as long-term property investments. Short-term finance was generally provided by the clearing or merchant banks in the form of loans secured against the site and sometimes the buildings. Long-term funding, often pre-arranged, was usually provided by insurance companies by way of fixed-interest mortgages. Occasionally, the development would not be retained by the developer but sold as an investment to an insurance company or directly to an occupier. This might be the case where the developer could not arrange suitable mortgage terms. It was rare for the financiers, with the exception of some merchant banks, to participate in the profit or risk of the development.

As inflation became a permanent feature of the British economy the insurance companies saw the disadvantages of granting fixed-interest mortgages and wanted to participate in the rental growth. At the same time, long-term interest rates rose and developers were faced with an initial shortfall of income over mortgage interest and capital repayments, often referred to as the 'reverse yield gap', which has remained an almost permanent feature of property financing (the gap narrowed and at times disappeared in the early 1990s). Thus insurance companies became less inclined to grant mortgages and developers were forced to give away some share of future rental growth in order to close the 'gap'. The insurance companies became more directly involved with the direct ownership of property and they were gradually joined by other financial institutions (see Table 4.1 and Figure 4.1). An increasingly active property investment market emerged and the traditional division of the roles began to blur.

At first, in order to attract the best investments, long-term investors began to compete with and take on the additional role of the short-term financiers. At the same time, some of the traditionally short-term financiers, the clearing banks and the merchant banks, began to seek a share in the equity of the development itself. As the competition for the best (the 'prime' investments) increased some of the insurance companies and pension funds – either on a project basis or by the acquisition of property companies – began to take on the role of the developer, accepting the additional element of risk in return for a marginally better long-term yield. The funding of developments on a long-term basis became dependent on the property satisfying the criteria of investors. Developers had a much wider choice of financial sources.

This level of activity increased to the height of the second post-war boom of 1971–73. Then, in late 1973 and early 1974, as a result of the rise in short-term interest rates, the rent freeze and the proposals for a first-letting tax, the property boom collapsed. Although Table 4.1 and Figure 4.1 shows that money continued to be invested by the institutions during that period, this was largely as a result of prior commitments. A more accurate reflection of the market conditions at that time is shown by the change in investment yields in Table 4.2. During this period virtually no new funds were made available for development and very few developers took on new schemes.

By the end of 1977 the market for completed and let investment had re-established itself firmly with yields returning to levels of 1971. The institutional investors emerged as the dominant force in prime commercial and industrial property markets during the first half of the 1980s. The criteria they adopted became increasingly narrow, leading to a widening of the gap between 'prime' or 'institutionally acceptable' properties and the rest. By 1981–82 prime investment yields had fallen to a historical low, as is shown in Table 4.2. Due to their selective approach, institutions kept yields low over a long period, despite rises and falls in the general level of interest rates. Many of the larger insurance companies began to carry out their own developments.

The banks became the dominant source of finance during the late 1980s development boom encouraged by the rapid increase in rents and capital values caused by occupier demand. They replaced the institutions who reduced their property investment in the mid-1980s (see Table 4.1), largely due to the better performance of other forms of investment such as equities (stocks and shares) compared with the poor performance of property in the early 1980s. In addition, developers preferred to obtain short-term 'debt' finance from the banks to enable them to sell their completed development into a rising market. Alternatively developers were able to secure medium-term loans or refinance initial short-term loans to enable them to retain their developments as investments. During the boom period in particular

Table 4.1 Net institutional investment by insurance companies and pension funds (1963–1993)

Year	Nominal (£m)	Real 1990 prices (£m)
1963	90	855
1964	91	822
1965	128	1110
1966	162	1351
1967	192	1571
1968	257	1990
1969	346	2543
1970	327	2239
1971	306	1914
1972	270	1568
1973	664	3497
1974	861	3836
1975	762	2709
1976	1056	3271
1977	948	2597
1978	1261	3192
1979	1264	2732
1980	1886	3536
1981	2040	3416
1982	2102	3315
1983	1518	2281
1984	1789	2563
1985	1402	1902
1986	1150	1510
1987	535	674
1988	1807	2139
1989	2094	2303
1990	587	587
1991	2171	2084
1992	1646	1534
1993	791	726

Net level of investment by insurance companies (life and other), pension funds, other funds, investment trusts and unit trusts. Source: PMA, CSO Economic Trends Financial Statistics.

Table 4.2 Prime property yields

Year	Offices (%)	Retail (%)	Industrial (%)
1970	6.25	6.50	8.25
1971	5.75	5.75	8.25
1972	4.00	4.25	7.00
1973	5.00	5.00	7.25
1974	8.00	8.00	9.50
1975	6.00	6.00	8.50
1976	6.50	6.25	8.25
1977	5.00	4.50	6.50
1978	4.75	4.25	6.50
1979	4.50	4.00	6.25
1980	4.50	4.00	6.25
1981	4.50	3.50	6.25
1982	4.75	3.75	6.75
1983	4.75	3.65	6.75
1984	4.75	3.65	6.75
1985	5.00	3.75	7.50
1986	5.00	4.00	8.00
1987	5.00	4.00	8.00
1988	5.00	4.00	7.50
1989	5.00	4.75	7.00
1990	5.50	5.25	8.50
1991	6.00	5.00	8.50
1992	6.60	5.20	8.70
1993	5.50	4.00	6.75
1994	4.75	4.25	6.75

Source: PMA, Healey and Baker Prime Yields.

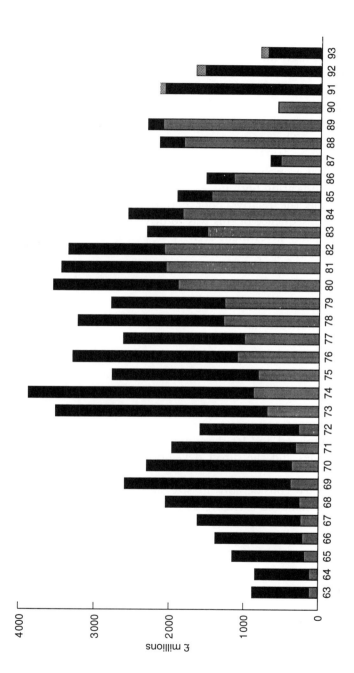

Figure 4.1 Institutional investment by insurance companies and pension funds (1963–1993). Solid bars: real 1990 prices. Dotted bars: nominal prices. Source: PMA, CSO Economic Trends Financial Statistics.

new financiers entered the commercial property market including foreign banks, foreign investors and to a lesser extent building societies.

The property market crashed in 1990 caused by a combination of the economic recession, high interest rates and an oversupply of new buildings. The banks emerged laden with property debt (see Section 4.2.3) and many development companies went into receivership or concentrated on their levels of debt. As Britain's economy has slowly improved since 1992–93 some development activity is taking place mainly on the basis of pre-lettings. The financial institutions re-entered the property investment market in 1993, encouraged by low property values, the expectation of rental growth (Table 4.1 and Figure 4.1 show net institutional investment indicating both increased selling and buying of property investments) and the relative poor performance of other forms of investment. The consequence of this was to drive yields down (see Table 4.2), although property investment activity slowed in 1994, largely due to the lack of availability of 'prime' stock.

We shall now examine each of the financial sources in detail and their influence on the property development process.

4.2.2   Financial institutions

Financial institution is the general term used in the property industry to describe pension funds, insurance companies, life assurance companies, investment trusts and unit trusts. They invest in property directly and indirectly through the ownership of shares in property investment companies and development companies. Direct property investment includes the owning of completed and let developments, the forward-funding of development schemes and the direct development of sites and existing properties.

Pension funds vary considerably in size and include the individual occupational funds managed exclusively for the employees of former/present nationalized industries and large publicly quoted companies, together with company and personal pension schemes managed by insurance and life assurance companies. They invest the premiums paid by the clients to achieve income and capital growth in real terms in order to meet the future payment obligations of pensioners on retirement. Many pension funds and schemes are under pressure due to the approaching maturity of their schemes (i.e. the ratio of expenditure on pensioners to income received from premiums is increasing), which is resulting in more emphasis on cash-flow and real income growth. In addition, many pension schemes are linked to the final salaries of employees which is adding to this pressure as incomes continue to rise above the rate of inflation. The Goode report and the government's White paper in 1994, on the reform of the pensions industry, recommends that funds should maintain a higher solvency (liquidity) ratio which may have future implications for the property industry.

Life assurance companies and insurance companies invest the premiums they receive on life and general insurance policies, respectively, to ensure long-term income growth to meet payment obligations when they occur.

Unit trusts are managed by financial institutions who offer investment management services, e.g. merchant banks. The unit trust will comprise unitholders such as small investors and institutions who are unable to take on the risk of direct investment. The trust will manage a portfolio of shares on behalf of the unitholders to obtain a spread of risk. There are two types of unit trust which specifically invest in property: authorized property unit trusts and unauthorized unit trusts. The former is strictly regulated to ensure that they invest in a diversified portfolio of low risk prime income producing property as part of a balanced portfolio, as their investors include private individuals. The latter are unregulated investing directly in a portfolio of properties and are attractive to tax exempt financial institutions such as charities and small pension funds.

All the financial institutions mentioned above tend to minimize risk and adopt a conservative approach to any investment as they are trustees of other people's money and are therefore are under constant pressure to perform. They invest in property as an alternative to other forms of investment such as stocks and shares (equities), together with bonds and gilts (fixed-interest income), both in the UK and throughout the world. By investing in different assets they give diversity to their investment portfolios and hence spread the risks involved. The extent to which a financial institution will invest in property largely depends on the size of the fund and the nature of their liabilities, and will vary according to the state of the economy and the performance for property investments relative to other investments. So although they are long-term investors they take a short-term view of performance, being strongly influenced by the recent performance of each type of asset, although they do forecast future trends (see Chapter 7).

At the start of the 1980s, when property investment by the financial institutions reached a peak (see Table 4.1 and Figure 4.1), property represented approximately 15% of all new investment, but declined to only 5% at the beginning of 1993 (DTZ Debenham Thorpe, 1993).

According to the latest survey by DTZ Debenham Thorpe (1994) the average weighting of property within a portfolio is around 5% in relation to pension fund assets and about 7% in relation to insurance company assets (includes life funds). The reduction of property investment in port-folios since the early 1980s has been partly due to the attraction of other forms of investment with equities both in the UK and abroad becoming dominant. There has been greater interest in the property market over the last 2 years (December 1994) partly due to the poor performance of other investments. However, as DTZ Debenham Thorpe found in their survey (1994) some of the smaller pension funds and life funds are mov-ing completely away from property investment while many of the larger pension funds and insurance companies are increasing their property investments.

We will now examine the advantages and disadvantages of investing in property from a financial institution's point of view. It is important to appreciate the factors that influence the investment decisions of financial institutions as this affects the funding and sale of completed develop-ment schemes.

(a) Hedge against inflation

One of the main reasons why the institutions first entered the property investment market in the 1960s was due to the fact that property repre-sented a 'hedge against inflation'. In other words rental growth out-stripped inflation and therefore represented an opportunity to achieve income gains in real terms. However, analysis of rental growth and infla-tion over the last 20 years shows that the concept of a 'hedge against inflation' is a myth. Research carried out by Richard Ellis (McIntosh, 1994) shows that from 1977 to 1986 (apart from the period 1980–1981) real rental growth in the property investment market was negative. With inflation at historically low levels, property investment performance exceeded 20% for some of 1994.

(b) Institutional lease

One of the principal advantages of property as against other assets is the existence of the 25 year institutional lease with upward-only rent reviews and 'privity of contract'.

Upward-only rent reviews ensure that typically every 5 years the rental income received by the investor rises or remains at the same level and cannot fall. However, this assumes that the tenant continues to pay the rent. The English legal doctrine of 'privity of contract' ensures that in the event of a default by any tenant (unless it is the original tenant) then the landlord has the ability to require the original tenant, followed by any assignees in turn, to pay the outstanding rent. Accordingly any tenant's contractual obligations will remain to the determination of the lease even if that tenant has assigned their interest in the lease. Tenants may assign their interest in the lease to another party provided that they have the prior approval of the landlord, which cannot be unreasonably withheld. Usually the test for such approval is the financial standing (covenant) of the tenant and typically institutions require the potential assignee to demonstrate that their last 3 year's trading profits exceed three times the rent payable. Accordingly, an investor in property is guaranteed an upward-only secure stream of income over 25 years, with some risk of voids due to tenant default.

However, the prevalence of the 25 year institutional lease has been under some pressure, particularly since the property slump in the early 1990s. There has been market pressure for shorter more flexible leases by tenants in line with the requirements of their businesses. In addition, the concepts of 'privity of contract' and upward-only rent reviews have been widely debated and cricitized, due to the recent experience of tenants in the economic recession. When the property market crashed in 1990 and the economy was in deep recession, market rents fell below rents agreed in the height of the development boom, but tenants could not benefit due to the upward-only rent review provisions in their leases. Also as many tenants went into receivership landlords were turning to original tenants to pay the rent under the 'privity of contract' provisions. The government were under some pressure from tenant pressure groups to legislate against upward-only rent reviews and privity of contract, but they have to decided to allow the introduction of a code of practice to be agreed between the interest groups representing landlords and tenants.

Another reason why property has traditionally formed part of an institution's investment portfolio is diversity of performance risk. If you examine total returns from property compared to equities and gilts (see Figure 4.2) it shows that property investment is not prone to short-term fluctuations, when compared to equities and gilts. In addition it shows that property investment performance behaves independently to equities and gilts. However, in their survey DTZ Debenham Thorpe (1993) found that about a quarter of funds thought the role of property in risk diversification was now largely irrelevant due to property's weak and more volatile performance since the mid-1980s.

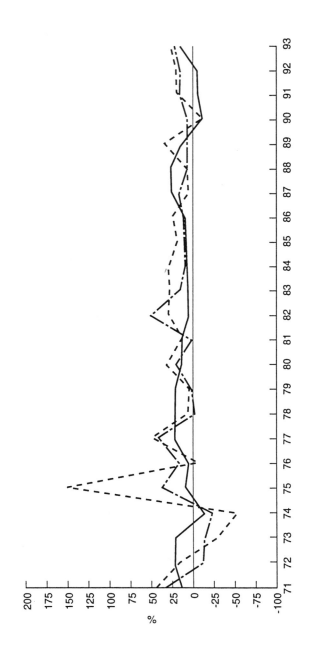

Figure 4.2 Comparison of property (solid line), equities (dashed line) and gilts (dot/dashed line) –percentage returns (1971–1993). Source: IPD Long Term Index – Investment Returns, IPD, WM, FT (pre–1974).

(c) Illiquidity and indivisibility

We turn now to the disadvantages of investment in property compared with other assets, the most significant being the illiquidity of property. As property represents a large investment in financial terms it cannot be sold quickly in response to market trends unlike equities. Selling a property can take months and it may not be possible to sell it if the market conditions are not favourable. Also the turnover of property sale transactions is low compared with other assets. Selling a property also involves high transaction costs such as agents' and solicitors' fees. Another significant problem linked to illiquidity is the indivisibility or 'lumpiness' of a property. This factor alone reduces the involvement of the smaller funds in the property investment market directly. The indivisibility and illiquidity of property has become the main focus of attention for those in the industry who are trying to improve the attractiveness of property as an investment. Later in this chapter (see Section 4.3.5) we will examine some of the attempts to introduce funding techniques to address this problem. It is widely believed that once illiquidity can be overcome a more active market property investment market may result, improving the attraction of property and involving some of the smaller funds in direct property ownership.

(d) Management

Property investment also involves a high degree of management measured in both time and cost. Although most management costs can be recovered from tenants under the terms of leases, it is still perceived as a disadvantage. Conversely active management of a property asset may improve the return received.

(e) Performance measurement

One of the main differences between property and other assets is the way in which performance is measured, making it difficult to compare property with other assets on a like for like basis. Property transactions take place in an imperfect market as transactions are conducted in private and not through a central market such as the Stock Exchange. Although, the flow of information on property transactions is improving all the time it is still imperfect (see Chapter 7 for details on the information sources available). The two main indicators of the state of the property market are rents and yields. The yield of a property investment is generally defined as the annual rental income received from a property expressed as a percentage of its purchase price or value (see Chapter 3). The yield is a measure of the property investor's perception of the future rental growth and capital growth against future risks, management expenses and illiquidity. Table 4.2 shows the movement of 'prime yields' since 1970. Prime yields are calculated by analysis of market transactions in 'prime' properties (sometimes referred to as 'institutionally acceptable' properties) which conform to the following narrow criteria:

1. Modern freehold or long leasehold property.
2. Good location.
3. Highest quality and specification.
4. Fully let and income producing to tenants with good covenants on fully repairing and insuring 25 year leases with frequent upward-only rent reviews.

Prime properties only represent a very small percentage of the entire property investment market. The movement of 'prime' yields represents a bench mark against which the yields of all properties can be measured. The movement of yields represents market sentiment about a particular type of property and the better the perceived prospects for either capital or rental growth, the lower the yield (higher capital values relative to income).

As we have already mentioned above in making a decision as to how much money to allocate to property purchases the financial institutions will have regard to the performance of both the property investment market and their existing portfolio of property. There are several established performance measurement indices which measure the performance of institutional property portfolios (see Chapter 7 for further details).

So far as expectations of future performance are concerned, institutions will typically hold monthly meetings to review forecasts of performance and reallocate funds to the different available asset classes. Property does not sit easily with such a rapid review timetable. The integration of property into a wider, multi-asset, context is generally made more difficult by the differences in terminology and valuation practice between property and other assets. In the past, this may have led to an unscientific and relatively static approach whereby, on a semi-permanent basis, a given level of a fund's assets were held in property. Some believe (although there is no conclusive evidence) that these issues of differing terminology and valuation practices may still result in there being lower exposures to property by some fund managers than otherwise would be the case: that property suffers prejudice in the investment allocation process.

The integration of property into wider multi-asset investment policy is now occurring as new techniques of valuation and portfolio construction are being applied to property. Fund managers like Prudential Portfolio Managers Ltd, examining the prospects for property in similar ways and using similar terminology, place property on a like for like basis with other asset classes.

Once a decision has been made on the allocation of money to property investment it is the responsibility of the property fund manager to make the decision on what type of property to purchase and in what location. Property investment policies are usually based on analysis by property type and region, looking at recent performance and future forecasts of growth (for further details on forecasting see Chapter 7). The policies and investment criteria of each institution will differ. However, they all tend to seek a balanced portfolio of property types, although the portfolio is usually weighted towards the property type which is performing well at that particular time. Most also try to spread their investments geographically, although office investment, in particular, will tend to be concentrated in London and the South East, but not exclusively. Most tend to adopt very rigid selection criteria when making decisions on what property investments to purchase. Many will only look for 'prime properties'. Accordingly, they will be looking at the best located properties of the highest quality, fully let on institutionally acceptable lease terms to tenants with good covenants. However, as properties falling within the definition of 'prime property' usually account for less than 10% of all properties at any one time institutions may have to compromise on some of the following factors:

1. Location
2. Quality of specification and design
3. Lease terms
4. Tenant

It may be impossible to obtain 'prime property' either because it is not available or the price being asked is too high (yield is too low) relative to the perception of future rental growth. Therefore, some may be willing to take a balanced view on a specific property analysing it on its own merits and adjust the yield they are prepared to accept to reflect the additional risk. Some may give different weight to each of the above factors depending on their investment requirements. For example, those funds who are concerned with the security of income rather than capital growth prospects will put greater emphasis on the quality of the tenant covenant. DTZ Debenham Thorpe found in their survey that some are adopting a 'back to basics' approach with more emphasis being placed on location and quality of building, rather than security of income, the rationale behind such emphasis being that rental growth will be stronger and the property will be easier to re-let affording more flexibility in relation to covenant strength and the length of the lease. The characteristics of a good location in relation to the various property types has already been examined in Chapter 2. Some institutions may be prepared to look at what may be considered secondary locations, where there is an under supply of quality stock, and where research shows there is strong prospects of rental growth.

The security of income factor in property investment is under some pressure as the outcome of the introduction of a code of practice on such matters as 'privity of contract' and rent reviews is awaited (see above). The institutions have already adapted over recent years to shorter leases, as leases of 15–20 years are more consistent with the economic life of a property. If in the future the majority of tenants resist 'upward-only' rent reviews and 'privity of contract' provisions in their leases then the institutions will need to reappraise their approach to property investment. Some consider it will mark the end of property as an investment whilst others argue it will bring property into line with other asset types. In reality tenants would have to pay higher rents for such flexible terms and yields would rise to reflect the risk of fluctuating incomes.

As well as purchasing completed and let developments as investments, many institutions carry out their own developments or provide development finance. Many will restrict their development activity to the redevelopment of properties in their own portfolio. Involvement in development, whether directly or indirectly, will depend on a particular institution's attitude to risk and their perception of the development cycle at any one particular time. Development is a riskier proposition than buying a standing investment. Currently there is renewed interest by the institutions in development and development funding – 55% of respondents in DTZ Debenham Thorpe's survey (1994) intended to start development activity in1994–95. This is due to a number of factors. Building costs and land values are comparatively low compared with the prices being currently sought for prime standing investments. In addition, development provides the institution with an opportunity to tailor a property in the absence of suitable stock available on the market. However, it is important to bear in mind that development represents a small proportion of all institutional property investment. A study by the Investment Property Databank (Darlow, 1989) between 1981 and 1988 revealed that expenditure by some 50 institutions (managing some 80 individual funds) on external development had averaged £131 million, equating to 16% of their total annual property investment. In support of this DTZ Debenham Thorpe's latest survey (1994) shows that institutional demand in 1995 is expected to be largely directed towards standing investments.

4.2.3 Banks and building societies

Banks who participate in the funding of development include both UK and foreign clearing and merchant banks. Banks during the 1980s became the dominant supplier of development finance as UK banks competed with foreign banks to fill the gap left by the financial institutions as they reduced their exposure to property. They extended their traditional role as short-term financiers and became involved in medium-term loans – usually up to 5 years beyond the completion of a development to coincide with the first rent review. They were encouraged by the dramatic growth in capital and rental values of property in the late 1980s property boom. Bank lending, as represented by outstanding loans to property companies, reached a peak in May 1991 at around £41 billion. During this time banks were also joined by some of the larger building societies (traditionally residential mortgage providers) although only on smaller schemes. Since the property crash of 1990 banks have been concentrating on reducing their debts outstanding to property companies, many of which are secured on the still unlet or unsold developments of the 1980s. There are currently signs that the banks are beginning to re-enter the property lending market albeit very conservatively, with a widespread avoidance of 'speculative' development funding, concentrating on lending on standing investments. They are tending to lend on prime low risk developments which are pre-let.

The banks are in business to make a direct financial gain from lending money and lending to property companies has always been viewed as profitable (subject to slumps). Bank lending may take the form of 'corporate' lending to a company by means of overdraft facilities or short-term loans. Alternatively, a loan may be made to enable a specific development project to proceed or for a developer to retain a development as an investment. We will examine the various loans available in Section 4.3.2. The banks will use the development or the investment property and/or the assets of the company as security for loans. Property is attractive as security for banks as it is a large identifiable asset with a resale value. However, as the experience of the late 1980s development boom and slump shows property values can fall. Accordingly, the banks' willingness to lend money to developers is determined by their confidence in the property market and the underlying economy at any particular time. It has been noted that UK banks have a tendency to follow their competitors when making decisions about their level of exposure to property (Gooby, 1992).

The clearing banks, due to their large deposit base, are the major providers of corporate finance loans to development and property companies. Some project finance is provided by the clearing banks on small schemes being developed by established customers. The merchant banks (some are subsidiaries of clearing banks) and specialist property lenders have smaller funds but have more property expertise. Therefore, they are more inclined to provide project loans and because of their expertise will take on more riskier loans (discussed below) in return for an equity stake in a project. On large development projects merchant banks may take on the role of 'lead' bank and put together a syndicate of banks to provide finance, and in doing so may underwrite the loan. Merchant banks also undertake investment management on behalf of institutional investors through investment funds and unit trusts. In practice the respective roles of both the clearing banks and merchant banks may overlap, particularly if they are associated. Many foreign banks are represented in the City of London, who operate in a similar way to British clearing banks.

There are also financial intermediaries who act as agents or financial advisers structuring development finance with banks and other sources for a developer in return for a fee, typically 1% of the value of the loan sought. Several of the large surveyor/agency firms such as Richard Ellis, Jones Lang Wotton and Chesterton have financial service arms.

Building Societies since the 1986 Building Societies Act have been allowed to provide corporate loans and loans secured on commercial rented property, provided such loans comprise less than 5% of total loans. However, many of the building societies had a bad experience in the latest boom and crash, particularly with residential developers, and have withdrawn from development funding. Other societies are still willing to fund commercial property investments. Typically they restrict themselves to small loans up to £10–15 million and tend to be less competitive than banks in relation to the interest rates they charge.

The banks' criteria for making loans will vary depending on the size of the development company, the nature and size of the development company, the nature and size of the development, the length of the loan, and the strength of the security being offered. In assessing the risk of corporate loans the bank will be concerned with the financial strength, property assets, track record, profits and cash-flow of the development company. In relation to loans on specific developments the banks will also be concerned with the security of the development project. The banks will wish to ensure that the property is well located, that the developer has the ability to complete the project and the scheme is viable. The in-house team of property experts with external advisers, if necessary, will carry out an assessment and valuation of the project. In the case of medium-term loans, where the developer wishes to retain the completed property until the first rent review, and loans on investment properties, the banks will also be concerned as to whether the rental income covers the interest payments. In the late 1980s banks were prepared to provide loans where the rental income did not cover the interest payments (referred to as deficit financing) on the basis that both rental and capital values were rising rapidly. This is no longer the case with the current upward trend of interest rates and low rental growth.

Banks have always tended to take a short-term view in relation to their lending policies, being concerned primarily with the underlying value of the development company and/or development project. As a result there has been widespread criticism of the banks' lending policies during the late 1980s with many laying the blame for the boom/slump at their door. The banks relied too heavily on the security and did not pay enough attention to assessing the risk attached to both borrower and projects. The banks in turn have been quick to blame the valuers whose opinions formed the basis for assessing loans evidenced by the many negligence cases pending against valuers. Both the banks and their advisers have learnt from their experience of the dramatic fall in property values in the early 1990s. This has led banks to review their policies and take a more medium-term view beyond prevailing market conditions. The Royal Institution of Chartered Surveyors are addressing the issue of valuations and in 1994 published the *Mallinson Report* recommending ways in which commercial valuations could be improved upon from the viewpoint of the client, which is the subject of much debate within the industry.

Property loans currently account for 10% (compared with over 12% at the peak in 1991–92) of all commercial lending by the banks (DTZ Debenham Thorpe, 1994) following writedowns and repayment of loans over the last few years. Merchant banks have particularly reduced their exposure to property. Banks have reduced their exposure to property in many ways. When property values slumped in 1990 many development companies were in default in relation to the conditions of both their corporate and project loans, as the value of their developments/properties were less than their outstanding debts. Banks have been willing to restructure many loans by a combination of measures such as renegotiating loan terms, refinancing loans, swapping 'debt' for 'equity' (converting part of the debt into mortgages) and forcing sales of assets. Many companies, particularly those with large development programmes, and specific projects went into receivership. Trader developers were particularly vulnerable and of those who remain many are being tightly controlled by their banks. Where the banks have decided to stand by a developer or project the problem of vacant or over-rented property remains. The institutional investment market is not interested in purchasing over-rented or secondary unlet property. Accordingly, the banks have had to make large write-downs over the last 4 years. Most of the medium-term loans negotiated during the late 1980s are now reaching maturity (1994–95). The question remains as to who is going to refinance the banks' debt in a market characterized with an oversupply of secondary space and low rental growth. Various ways are being examined to help the banks reduce the amount of debt, e.g. securitization, property bonds and debt for equity swaps.

4.2.4 Property companies and the stock market

As we have previously discussed in Chapter 1 there are two broad categories of company who participate in development: investors and traders. The investor type of company, usually referred to as a property company, is also a source of long-term finance as some purchase property investments for their portfolio as well as retaining their own developments. Their capacity to purchase property depends on their ability to raise finance. Property companies and development companies alike are partly financed by their own capital and that of their shareholders, and partly financed by borrowing money either short-term or long term. The level of 'gearing' (relationship between borrowed money and the company's own money) will vary between companies. Property companies as opposed to 'trader' development companies tend to have a lower level of gearing due to the strength of their asset base.

Property companies vary from small private firms to large publicly quoted companies. According to DTZ Debenham Thorpe (1993) the largest 10 companies account for over 75% of the quoted property sector's capital value. Some specialize in a particular geographical location, while others hold large portfolios of a variety of property types in both the UK and overseas. Their prime objective is to make a direct profit from their investment and development activities. Some will take a longer-term view than others to the extent to which they 'trade on' their investments and completed developments. They have a responsibility to their shareholders to maintain the share price and provide dividends. Property companies view property investment as both a source of income and as an asset providing security for borrowed money. Property companies, particularly the large quoted ones, will tend to concentrate their investment activities on prime and good secondary properties. However, unlike the institutions, they see the management aspects of property investment as an advantage. They have both the management and development skills in-house to improve the value of properties. Also they are not adverse to multi-let properties provided they are in a good location and of high quality. They may purchase investment properties which are not fully let or are nearing the end of the lease with redevelopment potential.

The shareholders of property companies are a combination of financial institutions and private individuals. Financial institutions invest in property company shares instead of or in addition to their direct property investments. However, it is not tax efficient for pension funds (who are non-tax payers) to invest in property company shares when compared with investing in property directly. This is because corporation tax is paid on the company's profits before dividends on the shares are paid and capital gains tax is paid on property sales. Tax paying shareholders will be taxed on the dividends and on any capital gains from selling the shares.

With a quoted property investment company, the shares are valued by the Stock Market below the value of the assets of the company attributable to the shares. This is known as the NAV or net asset value per share. The long-term historic discount rate is 20% (DTZ Debenham Thorpe, 1994). This discount to asset value is due to the tax disadvantage of the company as capital gains tax might be payable on the sale of their assets. The amount to which shares are discounted varies with stock market conditions and the state of the property market. However, the value of property company shares fluctuates more widely than the value of property, regardless of the state of the property market. This was shown to be the case on 'Black Monday' in October 1987, when the property market was booming but the shares of property companies fell dramatically along with all other equities, although when viewed over the longer term there is a correlation between the trends in property values and the share prices of property companies (Rodney and Rydin, 1989). Other factors affecting the value of the share price of property companies include the financial strength of the company including its level of gearing as evidenced by their balance sheet and the perceived strength of the management team. The price earnings ratio (P/E) is the main yardstick used to assess property trading companies.

Equity finance can be raised by issuing various forms of shares in a company, with investors directly participating in the profits and risks of the company. New property companies may float on the Stock Market and raise money by selling shares. Quoted companies can issue new ordinary shares or preference shares to raise equity finance for their development activities, depending on Stock Market conditions, the overall performance of property company shares and the NAV per share of a particular company (see Section 4.3.4 for further details). Such finance may also be used to repay bank borrowings and other debts or retain certain developments in an investment portfolio. In addition, companies can also raise debt finance via various methods on the Stock Market. Long-term debt finance is capital borrowed from investors which usually involves fixed interest, and may be secured on the company or unsecured. Debt finance usually has to be repaid by a certain date or converted into shares (equity). Debt finance instruments became popular during the late 1980s as an alternative means of providing long-term finance to hold developments as investments.

We shall examine the methods whereby property companies can raise money in the form of debt or equity on the Stock Market in further detail in Section 4.3.4.

4.2.5   Overseas investors

Since 1988 overseas investors have become significant participants in the property investment market, particularly in central London. Investment by overseas investors reached a peak in 1989–90 at over £3 billion in each year, exceeding the investment by UK institutions. It dropped to around £2 billion a year in 1993–94, but between 1988 and 1994 it has accounted for 15% of all UK commercial property transactions (DTZ Debenham Thorpe, 1994). The UK property market is attractive to overseas investors due to the unique existence of the 'institutional' lease offering security of income for up to 25 years.

The source of overseas investment has varied since 1988. The Japanese and the Scandinavians led the way during the boom period between 1987 and 1990. The UK government's favourable treatment of foreign investment together with the lifting of Japanese government's restrictions provided the impetus for Japanese investors, developers and contractors. The lifting of restrictions by the Swedish government on overseas investment by their property companies and life funds led to the Scandinavian interest. They were both attracted by the performance of the UK's economy and its relative stability at that time. Their development companies became involved in direct development either in partnership or on their own account. However, many have since gone into receivership following the collapse in the market in 1990.

European investors, particularly the Germans, became significant since 1992 due to the relative performance of the UK economy against their own economies and the continuing deregulation of cross-border investment by the European Union. They were joined by American, Middle Eastern and Far Eastern investors in the years of 1993 and 1994. All have tended, in contrast to the Japanese and the Scandinavians earlier, to invest in standing investments rather than developments. Like the UK institutions they are interested in 'prime' properties let on institutional leases to good covenants.

Overseas investors have tended to concentrate their investment activities in central London offices, but DTZ Debenham Thorpe (1994) have found that as they become more familiar with the UK they are diversifying their purchases and looking outside London. In addition overseas investors are more likely to invest in larger properties worth over £50 million, the upper limit of all but a few of the UK institutions.

4.2.6 Private individuals

Private individuals do purchase property investments at the lower end of the market. They tend to concentrate their purchases on secondary and tertiary commercial property with high yields. It is not possible to quantify the extent of their involvement as there are no research figures. They tend to be precluded from the 'prime' market due to the significant sums of money involved and the existence of the 'reverse yield' gap. Private individuals are attracted to high yielding properties as income will very often be in excess of interest rates. However, participation in the lower end of the property market is very risky, involving intensive management and regular voids.

4.2.7 Joint venture partners

A development company may raise finance or secure the acquisition of land by forming a partnership or a joint venture company with a third party to carry out a specific development or a whole series of development projects. The basic principle behind forming a partnership from the developer's point of view is to secure either finance or land in return for a share in the profits of the development scheme or the joint venture company. The joint share given to the third party will depend largely on the value of their contribution combined with the extent they wish to participate in the risk of the scheme.

There are many forms and methods of forming partnerships or joint ventures for the purpose of carrying out property development. It is beyond the scope of this book, as an introductory text, to examine them in detail. However, we will briefly examine the reasons behind forming partnerships and joint ventures, with some examples.

A partnership may involve any combination of sharing the risks and rewards of a scheme via many different contractual and company arrangements. In addition the partners to a scheme may take an active or passive role in the scheme. Tax and financial considerations may determine to a large extent the formal structure of the partnership arrangement. A joint venture may also take many forms, but in its 'purest' form the parties participate in the development and distribute the profits in equal shares usually by forming a joint company.

Developers are typically reluctant to share profits with third parties, unless it is the only way of securing a particular site or finance for a development scheme. Partnerships with landowners may be required if the landowner wishes to participate in the profits of the development scheme or wishes to retain a long-term legal interest in the property preferring income to a one off capital receipt. Local authorities and other public bodies, e.g. British Rail (now Railtrack), may only grant long leasehold interests to developers due to their need to retain a continuing interest in the property for financial or operational reasons. We have discussed the restrictions placed on local authorities in relation to capital receipts and the forming of joint venture companies. The ground rent and profit sharing arrangements will be determined by the amount of risk the landowner wishes to bear.

If a developer is involved with a particularly large or complex development then the spreading of the risk through partnership arrangements is the only prudent way of undertaking such schemes, which will often be beyond the financial capability of all but the very largest companies. With such schemes one or more partners may be involved and may include the landowner, contractor or another development company. Developers may also form joint venture companies with other development companies who may have the expertise or experience required for a particular type of development which is seen as vital to the success of the scheme. There are also examples of developers forming joint venture companies with retailers to combine their respective experience and market knowledge.

During the late 1980s joint venture companies were formed on some of the larger developments to enable the partners to arrange 'limited' or 'non-recourse' finance off-balance sheet so the borrowings did not appear on either of the partner's respective balance sheets. The rules are tightening up on the these arrangements as we will see later in this chapter.

Whatever the reasons for forming a partnership to finance a scheme the developer must ensure that the definition of the profit is clearly set out. In the case of a joint venture company the profit will be distributed through the company accounts. A developer must rehearse every possible outcome of the scheme to ensure that any partnership arrangement will work and the true intentions of the parties have been carefully documented.

4.2.8   Government assistance

As we have already examined in Chapter 2 (see Section 2.6.2) there are grants available from the government to developers directly or through a partnership with local authorities on a City Challenge scheme. In order to qualify for grant assistance the developer's proposal must be in a specifically designated area – UDC area, UPA or designated local authority areas in Scotland and Wales – in order to qualify for grant assistance; although other areas qualify for assistance depending on the particular circumstances. The two main grants currently available are City Grant (plus similar grants in Scotland and Wales) and Derelict Land Grant but they are currently being reviewed by English Partnerships. In order to qualify for assistance a developer must prove that the scheme would not proceed without grant assistance as the development costs exceed the development value. Furthermore the development proposal must not just involve physical regeneration but provide jobs relevant to the local community. The indications from English Partnerships are that more innovative methods of providing assistance to developers will be preferred in future to just handing out grants. For example, UDCs are now able to offer rent guarantees. There may be a move towards more equity participation through the provision of land and loans to encourage development in urban regeneration areas in recognition of the fact that funding is difficult to achieve. Developers working in partnership with local authorities may also bid for City Challenge funding in UPAs.

Investors are encouraged by the government to invest in schemes in areas designated as Enterprise Zones with 100% tax allowances on investments by individuals and companies with relief at the investor's highest rate of tax (see Section 2.6.3).

4.3   METHODS OF DEVELOPMENT FINANCE

As we have seen there are a number of different sources for development finance. There are many methods of obtaining development finance from the above sources and the 1980s saw increasingly innovative techniques applied. We shall now examine the various well-established methods and briefly look at some newer forms of finance which have emerged during the late 1980s with varying degrees of success.

The choice of both source and method of development finance will depend on how much equity (the developer's own capital) that the developer is able and willing to commit to a scheme. If the developer has insufficient capital then the aim is to arrange as much external finance as possible, whilst retaining as much of the equity as possible without giving away bank or personal guarantees. A decision has to be made as to how much risk the developer wishes to pass on to the financier in return for a share in the financial success of the scheme. The availability and choice of finance will depend largely on the company's size, financial strength, track record, the characteristics of the development scheme to be funded and the duration of the scheme. Whatever method is chosen the developer will always need to be fully aware of the tax consequences.

4.3.1   Forward-funding with an institution

Forward-funding is the term given to the method of development finance which involves a pension fund or insurance company agreeing to provide short-term development finance and to purchase the completed property as an investment. This happens at the start or, at least, at an early stage in the development process. This method of finance reduces the developer's exposure to risk, accordingly the terms usually agreed with the institution reflect this. From the institution's point of view, this method of acquiring a property investment has several advantages over purchasing a ready-made investment. It provides the institution with a slightly higher yield than a ready-made investment, reflecting the slightly greater risk. By being involved in the development process, the institution influences the design of the scheme and the choice of tenant. In addition, if there is a rise in rents during the development period, the institution shares in the rental growth.

As already discussed (Section 4.2.2), the proposed development must fall into the 'prime' category if the developer is to be successful in securing forward-funding, although if the proposed scheme is very large in terms of its lot size (in practice, over £50 million), then the number of funds in the market is reduced. Therefore, it is essential for developers to consider this fact when purchasing a site as it will affect the way in which they evaluate the development opportunity.

On the assumption that the developer is purchasing, or has just purchased, a site in a prime location with the benefit of a planning consent, then the next step is to approach the institutions direct or through their agents. Agents have an important role to play in the forward-funding of a scheme as they have a good knowledge of the institutional investment market. In addition, because so many of them are retained by institutions themselves they have good contacts. The developer may also have established a good working relationship with particular institutions they may have worked with in the past. Institutions themselves will tend to adopt a proactive approach and directly or through agents seek out the development opportunities themselves in accordance with their investment criteria. They will have identified through their research the property type and location they are interested in.

The developer will usually prepare a brochure for presentation to those institutions who express initial interest in the scheme. The brochure will typically describe the nature and location of the development, with supporting illustrative material – planning consents, site investigation reports and specifications. Developers will need to sell their track record and experience on similar schemes. They will also need to provide an analysis of the market in terms of the balance between supply and demand (see Chapter 7) for similar schemes. An initial appraisal of the scheme will also be included as a starting point for negotiations on the value and cost of the proposed scheme, with any supporting evidence such as cost plans. However, the institution will not rely on either the market analysis or appraisal by the developer and will carry out their own evaluation to assess the risk, using a variety of sensitivity techniques (see Chapter 3).

The institution, having decided the proposition fits within its investment criteria, will need to satisfy itself that the proposed development is viable, that there is a demand for the development, that the specification and design of the building is of the highest quality, and that the developer has a satisfactory track record and expertise. The developer may need to be able to guarantee the investor's return at the end of the development period depending on the terms negotiated, so the institution will need to examine the developer's financial standing. The institution will ask itself the question is the developer going to produce the scheme on budget and within time?

Once a particular fund has agreed in principle to the forward-funding of the developer's scheme then negotiations take place over the financial aspects of the agreement. There are various types of arrangements that can be entered into largely depending on the state of the market, the nature of the scheme and the financial standing of the developer. Funds will tend to tailor the arrangements to suit the particular development and their view of the market. There are many characteristics which are typical of most deals and the variation between arrangements will be reflected in the balance of risk and reward between the parties.

We shall now examine each of the typical elements of a funding agreement:

(a) Yield

The fund and the developer will agree at the outset the appropriate yield (and therefore capitalization rate) applicable to the scheme. The yield will usually be agreed on the basis of market evidence and the fund's perception of the risk involved. As we have already discussed in Chapter 3 the yield is a measure of the institution's perception of risk weighed up against the rental growth prospects. In forward-funding arrangements yields will normally be discounted by around 1–2% (i.e. 1–2% higher) from the market yield for standing investments to reflect the additional risk the institution is taking by participating in the development process. The yield will be fixed at the agreed level.

(b) Rent

In some cases what is known as a 'base rent' will be agreed which will be based on market evidence of recent lettings of similar properties in the area. Obviously, if there is a pre-letting in place then the rent will be agreed with the tenant in the 'agreement to lease', although there may be provisions for a review on completion of the scheme. If the rent achieved on the scheme exceeds the base rate then there is normally a provision to share the benefit of this. This element of the rental income is known as the 'overage' and is usually shared by the fund and the developer. The fund may cap the 'overage' rent at a certain level because the developer will be motivated to achieve the highest rent and the fund will wish to safeguard against 'over-renting' the property which would be detrimental to rental growth prospects in the future.

(c) Costs

The developer will present to the fund a detailed estimate of development costs. These will be analysed by the fund's in-house building surveyors or externally appointed consultants. The institution will usually wish to cap the total development costs at a certain level, allowing for interest. The developer will be under an obligation not to exceed these costs. If the maximum agreed limit is exceeded then the developer will be responsible for funding the balance. There may be a provision within the agreed development cost for the developer's own internal costs such as project management fees and overheads. The maximum agreed development cost will typically relate to previously agreed plans and specifications. The developer will be under an obligation not to vary either the plans or specifications without the prior approval of the fund. There may be a provision which enables a variation of the agreed costs due to agreed variations.

The fund will provide the short-term finance at an interest rate to reflect their opportunity cost of money and not the cost of borrowing, so in practice the rate has tended to average 7–9%. This is because institutions do not need to borrow money, but regard the provision of short-term development finance as part of their investment. Moneys will be advanced to the developer on the production of architect's certificates in respect of the building costs and invoices in respect of all other costs. Interest will accrue and be rolled up until practical completion or until the scheme is fully let depending on whether the developer is responsible for any shortfall in rent. At the same time the developer will receive any profit, calculated as the development value less the development costs advanced in accordance with the terms of the funding agreement, although the fund will keep a retention equivalent or greater than that agreed with the building contractor, depending on the existence of any defects or work outstanding.

Depending on market conditions, the developer may be able to secure a profit on the value of the land if the value of the land at the time of the funding agreement is greater than the cost of acquisition.

(d) Developer's profit

The calculation of the developer's profit will depend on the type of funding arrangement entered into. The developer and fund will agree either a base rent or a priority yield method of funding.

(i) 'Base rent' arrangement   On completion of a scheme based on a 'base rent' arrangement the total development cost (including interest) up to any maximum agreed limit will be deducted from the agreed net development value for the scheme. The net development value is the total rent achieved up to the agreed base rent multiplied by the agreed Year's Purchase (a reciprocal of the yield) less the institution's costs of purchase. The balance of the calculation will represent the developer's profit, often referred to as a 'balancing payment'.

As an example, we will look at the evaluation in Chapter 2 on the assumption of a forward-funded deal with an agreed base rent. Assume that the developer has agreed with the institution a base rental value of £688 500 per annum, based on a net lettable area of 38 250 sq ft (3553 sq m) at a base rent of £18 p.s.f. (£193.75 p.s.m.) and a yield of 7.5%. It is also agreed that any rent achieved above the base rent will be split 50% to the developer and 50% to the institution. If the rent achieved is £775 000 per annum and the development cost is £8 000 000, then the developer's profit is calculated in the following way:

Example 4.1

Base rent plus	£688 500 p.a.
50% overage (i.e. £775 000 p.a. – £688 500 p.a. divided by 2)	£43 250 p.a.
	£731 750 p.a.
Capitalized at 7.5%	13.33
Gross Development Value	£9 754 228
Less purchaser's costs @ 2.75% equals	£9 493 166
Less development cost	£8 000 000
Balancing payment to developer (i.e. developer's profit)	£1 493 166
Developer's profit as percentage of cost	18.7%

The fund's profit is represented by the movement in the initial yield from 7.5 to 8.2% calculated by dividing the rent achieved by the development value and multiplying by 100, i.e. (£775 000/£9 493 166 × 100 = 8.2%).

(ii) 'Priority yield' arrangement   A 'priority yield' arrangement provides the fund with the first or 'priority' slice of the rental income before the developer takes a profit, providing a guaranteed return yield on the institution's investment. The 'priority yield' is particularly used where the costs or rents are perceived to be subject to a greater uncertainty (e.g. with lengthy schemes) and therefore present a greater risk.

Alternatively, the fund will agree with the developer the priority yield – in the example below 7.5% – and will receive as a priority slice 7.5% of the development costs. Then the developer will receive as the next slice an agreed percentage of the development cost – in this example 1% – which is then capitalized at the agreed base yield. This equates to the developer's required profit as agreed with the fund. The remaining rental income is split 50/50 or as otherwise agreed, then capitalized at the agreed base yield. Again using the evaluation in Chapter 2 we now re-work Example 4.1 on the basis of a 'priority yield' arrangement. The developer's profit is calculated in the following way:

Example 4.2

Development cost	£8 000 000
Achieved rental income	£775 000 p.a.
Fund receives first slice of rental income	
@ 7.5% of development cost	£600 000
Developer receives next slice of rental income	
@ 1% of development cost	£80 000
Developer/fund share balance of rental income 50%/50%	
(i.e. £95 000 divided by 2)	£47 000
Developer's profit is share of rental achieved	£127 500
Capitalized at 7.5%	13.33
Developer's profit	£1 699 575
Developer's profit as % cost	21.24%

The fund's profit is represented by the movement in the initial yield from 7.5 to 8.2% as in Example 4.1.

(e) Developer's guarantees and performance obligations

In addition to controlling their maximum funding commitment by capping costs the fund may require certain guarantees from the developer. Unless the scheme is entirely pre-let then the fund may require the developer to guarantee any shortfall in rental income until the scheme is fully let or 3–5 years after completion whichever is the earlier.

Alternatively, the fund may require the developer to enter into a short-term lease for 3–5 years after completion. Bank guarantees or parent company guarantees may also be required to support the potential rent liability to meet any costs exceeding the agreed limit.

If the developer wishes not to provide such guarantees or enter into a lease then a 'profit erosion' arrangement may be entered into. This is described in further detail in the case study at the end of the chapter. Under this arrangement interest and costs will continue to accrue until the scheme is income producing or 3 years after completion whichever is the earlier. At this point any profit due to the developer will be calculated and may have been entirely eroded through an increase in the development costs above any agreed limit or a decrease in the rent actually achieved. However, the developer will then be able to completely walk away from the development without any further commitments.

Typical of most funding arrangements is an obligation by the developer to perform – to build the scheme in accordance with the agreed specification and plans within the time and budget agreed. In addition, the developer has to ensure that the professional team performs and that collateral warranties are procured for the benefit of the fund (see Chapter 6 for further details on collateral warranties). Throughout the development the fund's interest will be protected by their surveyor who will oversee the project, attending site meetings as an observer, to ensure the developer is performing in accordance with the agreed specification and plans.

(f) Lettings

The developer will need to obtain the approval of the fund to all lettings. The funding agreement will usually specify on what terms the fund will be prepared to grant leases – a standard form is usually attached to the funding agreement. There may be a provision which allows only 25 year leases with upward-only rent reviews. However, increasingly more flexible lease lengths are allowed with or without breaks although 15 years is usually the minimum lease length acceptable unless circumstances dictate otherwise. In addition, the agreement will specify an acceptable tenant. A typical arrangement may specify that the tenant's profits for the last 3 years must exceed a sum three times the rent or total liability (including service charges). The fund may specify that only single lettings are acceptable, although floor by floor lettings might be acceptable after a certain length of time from practical completion.

(g) Sale and leaseback

As an alternative to the above described arrangements a sale and lease-back arrangement may be entered into whereby the developer retains an interest in the investment created by the development. These type of arrangements typified the funding deals arranged in the 1970s and early 1980s. They are the exception today as institutions prefer to retain total control. A sale and leaseback involves the freehold of the scheme passing to the fund on completion with the fund simultaneously granting a long lease to the developer, who in turn grants a sublease to an occupational tenant. There are many variations to this arrangement depending on the method of sharing the rental income. Sale and leaseback arrangements may be either 'top sliced' or 'vertically sliced'. With the 'top slice' arrangement the fund receives rent from the developer in accordance with the required yield. The developer is then able to retain any profit from letting the property at a higher rent than that payable to the fund. However, with upward-only rent reviews in the developer's lease with the fund the developer's profit rent may be rapidly eroded over time. This means the developer's interest is only saleable to the fund. Therefore, the 'vertical slice' arrangement is better from the developer's viewpoint as the fund and the developer share the rental income from the property in relation to an initially agreed percentage throughout the length of the lease.

Institutions are more likely to enter into sale and leaseback arrangements directly with the occupiers to create attractive property investments with tenants of good financial standing.

In the current market conditions (December 1994), where there is an oversupply of space (mainly secondary) many institutions are only entering forward-funding deals on the basis of a pre-letting. Alternatively, they are prepared to enter into arrangements with a developer of good financial standing where the rent is guaranteed for 3–5 years by the developer. However, forward-funding deals are currently being achieved on the basis of speculative schemes provided there is a lack of 'prime' space on the market and there is proven demand, where the risk is quantifiable by the institution. An example of such a deal is contained in the case study at the end of this chapter.

4.3.2 Bank loans

As we have already discussed, the clearing and merchant banks provide short-term development finance, either on a 'rolling' or project-by-project basis, by means of overdraft facilities or short-term corporate loans secured against the assets of the development company or project loans secured against a particular development. With the dramatic increase in bank lending over the 1980s, various different methods of bank lending were introduced and development companies sought bank finance beyond the construction period up to the first rent review.

For many developers, especially the smaller ones, forward-funding is difficult to obtain as they are unable to provide the requested guarantees. Also 'prime' properties which are acceptable to institutions represent a very small part of the market and to some extent tend to be geographically restricted to London and towns in the South East. Large development projects, typical of the late 1980s, are beyond the capacity of all but a few of the larger funds. For instance, if a development company is seeking to forward fund a retail town centre scheme worth in excess of £100 million, then they may need to involve two funds and the choice is very limited. From the developer's point of view, borrowing from a bank allows greater flexibility and enables the developer to benefit from all of the growth, unless some of the equity is given away. The developer can repay or refinance the debt when the time is right and sell on the completed investment at a higher price. In addition, the developer will not be subject to the same degree of supervision through the development process. In the market conditions that prevailed in the late 1980s when rents and capital values rose rapidly it was more profitable for developers to arrange debt finance as opposed to equity finance for the reasons mentioned above.

(a) Corporate loans

As we have already discussed previously (Section 4.2.3), development companies can arrange overdraft facilities or loan facilities with clearing banks secured on their assets. With corporate lending the bank is concerned with the strength of the company, its assets, profits and cash-flow. Accordingly, obtaining bank loans in this way is more appropriate to investor-developers and large developers rather than trader-developers and smaller developers as they have large asset bases which can provide the necessary security for bank borrowings. Corporate loans can be obtained at lower interest rates than project loans.

(b) Project loans

Alternatively, development companies can arrange project loans which are secured against a specific development project. Banks normally provide loans which represent 65–70% of the development value or 70–80% of the development cost. Developers have to provide the balance of moneys required from their own resources. The banks limit their total loan to allow for the risk of a reduction in value of the scheme during the period of the loan. In addition, by insisting on an equity injection by the developer they are committing the developer. This equity provision is normally required at the outset of the development to motivate the developer to complete the scheme. Also the developer is totally responsible for any cost over-runs. The loan to value ratio depends on the risk perceived by the bank. Obviously a pre-let or pre-sold development represents less risk than a totally speculative one. During the late 1980s it was possible to secure between 85 and 100% of development costs through various layers of bank finance using a combination of insurance, 'mezzanine' finance and a profit-sharing arrangements. Lending conditions will vary according to the banking sector's confidence in the property market.

Project loans are attractive to the smaller trading companies as they are not worth enough to fund their full development programme through corporate loans. For many larger companies, these loans can be carried off the parent company's balance sheet by forming joint ventures with the bank in a subsidiary company. The borrowing associated with the property would not appear on the parent company's balance sheet. This enabled development companies in the 1980s to have development programmes which would have otherwise been impossible as 'gearing' would have increased to an unacceptable level ('gearing' is the relationship between borrowed money and the company's own money). However, the regulations regarding such off-balance sheet subsidiaries are currently being reviewed and tightened.

When the banks lend against a particular development project, it will form all or part of the security for the loan. The developer needs to provide similar information to the bank as to a funding institution. The banks will wish to ensure that the property is well located, that the developer has the ability to complete the project and the scheme is viable. This requires the bank to have knowledge of the property market either through in-house staff or external advice from firms of chartered surveyors in order to assess the risk involved. The developer will need to present the proposal to the bank in the form of a package very similar to that required by a funding institution (see Section 4.3.1).

However, as the bank is viewing the scheme as a form of security and not investment, it is more concerned with the underlying value of the scheme rather than the details of the specification. The appraisal, therefore, forms the most important part of the presentation together with all the supporting information and market analysis. Equally important is the track record of the developer in carrying out similar schemes. The banks will also examine the financial strength of the development company although this may not form part of the security of the loan.

The bank will employ either its in-house team of experts or external surveyors/valuers to report on the proposal and provide a valuation of the proposed scheme. Part of the process will be an analysis of the risks involved and this should be reflected in the terms offered to the developer. The bank will be concerned to ensure there is sufficiently contingency and profit built into the appraisal to provide a sufficient margin for cost increases during the period of the loan.

Another element of the risk involved in bank loans is represented by fluctuations in interest rates which the bank may be concerned to limit (see below). Interest rates on bank loans are normally fixed but only in relation to the base rate or LIBOR (the London Interbank Offered Rate, i.e. the rate of interest between banks). It is agreed between the bank and the developer that the interest rate is, say, 1.5% above LIBOR. Generally interest rates will be higher on project finance loans than corporate loans due to the increased risk that the banks face. The interest rate margin may be less if the developer pre-lets or pre-sells the property before completion as this substantially reduces the bank's risk. Also the interest-rate margin on an investment loan would be lower than on a speculative development loan. Interest rates on short-term loans are likely to be floating whereas on long-term loans they are more likely to be fixed.

The bank should also consider how the loan will be repaid either by the sale of the completed scheme or by refinancing from another bank. It must be remembered that on completion of the scheme the rental income will usually be insufficient to cover the interest costs on the loan, due to the 'reverse yield' gap problem where yields on property investment tend to be lower than medium to long-term interest rates.

The bank will also need to protect its 'security' by obtaining a first legal charge on the site and development. In the event of default on the loan the bank will be able to obtain ownership of the development if required. It may also require a floating charge over the assets of the development company. The bank will need to be able to be legally capable of stepping into the 'developer's shoes' in the event of any default which may require the legal assignment of the building contract and any pre-sale or pre-letting agreements. Like a funding institution the bank will also require collateral warranties from the professional team.

Guarantees may be required from the parent company or a third party, if the financial strength of the development company is not considered adequate. A full recourse loan will involve the parent company providing a full guarantee on the developer's capital and interest payments together with a guarantee that the project will be completed.

During the 1980s limited and non-recourse loans became an attractive proposition for developers wishing to finance their development projects whilst providing limited or no guarantees. With a 'limited recourse' loan the parent company may only have to guarantee cost and interest over-runs. Limited recourse loans were normally granted for the construction period of a project and up until first rent review. A 'non-recourse' loan involves no guarantee with the only security for the bank being the development project itself. However, in practice the parent company is still responsible and it would be very difficult for a developer to simply walk away from the scheme without damaging their reputation.

The developer needs to take account of the considerable costs involved in bank finance which are usually up-front. The developer will need to pay for the cost of carrying out their own appraisal and presentation to the bank. In addition, there are several fees payable to the bank. An arrangement fee will be charged by the bank to cover the cost of carrying out a valuation and assessment of the project. This fee may include an element of profit depending on the risks involved. There will also be a management fee to cover the bank's costs in monitoring the project consisting of mainly surveyor's fees. Such charges can represent up to 3–10% of the value of the loan. There may be a 'non-utilization' or 'commitment' fee on the part of the loan not drawn down initially as the bank will have to retain the full loan facility and cannot commit the funds elsewhere.

There are some variations on the basic project loan described above.

(i) **Investment loans**   Development companies, wishing to retain a development, can secure the option to convert the project loan into an investment loan on the completion of the project once it is let, usually up until the first rent review. Alternatively, the developer may agree a combined project and investment loan from the outset. Also, the developer may be able to refinance a loan on completion on better terms than a previous short-term development loan. On investment loans banks will normally lend up to three quarters of the value but there is always the problem of the rent not covering the interest payments. Banks wish to see the interest covered by the rental income and, therefore, can limit the loan. This may be relaxed where the property is reversionary (let below market value) or the parent company guarantees the shortfall interest. Otherwise the banks may require the developer to cap the interest rate or re-arrange the payments on the loan (see below). The interest rates on investment loans are usually lower than on far riskier speculative project loans and the risk to the bank will depend on the financial standing of the tenant.

(ii) 'Mezzanine' finance   A project loan may be split into different layers known as 'senior debt and 'mezzanine' debt if the developer is unable to provide the normal equity requirement or wishes to increase the amount of the loan above the normal loan to cost ratios. The 'senior' debt usually represents the first 70% of the cost of the development scheme like a straightforward project loan. 'Senior' debt is usually provided by the major UK clearing banks. This debt may represent more than 70% if the development is pre-let or pre-sold. When a developer wants to borrow more of the cost of the project than 70%, additional money may be raised in the form of 'mezzanine' finance. Also the bank may increase the 'senior' debt exposure to 85% of cost with a commercial mortgage indemnity scheme. Mezzanine finance may involve the developer losing some of the equity, or alternatively, taking out an insurance policy. This mezzanine element is normally provided by merchant banks and specialist property lenders. As this mezzanine level of finance is more risky, the bank will charge a higher interest rate (usually 1–2% higher than the rate on the senior debt) or require a share in the profits of the development. The banks will also require full guarantees from the parent company.

If a mortgage indemnity insurance policy is taken out it will reimburse the lender if the loan is not repaid in full and will involve the developer paying a substantial one-off premium to a specialist insurance company. The policy may cover the mezzanine layer of the loan or the entire loan. Equity sharing with the bank may involve a profit share or an option on a legal interest in the scheme. Those banks willing to participate in the equity of the scheme are limited to those with sufficient property expertise. Very often due to the tax complications the profit share will be expressed as a fee. In this instance, the bank will become part of the development team and become involved in the decision-making process.

(iii) Syndicated project loans   The necessary development finance may need to be borrowed from more than one bank. In particular, larger loans, say, in excess of £20 million, are more likely to be syndicated among a group of banks by a 'lead' or agent bank. In the late 1980s many of the large office projects in the City of London were financed in this way. Each bank shares in proportion the risk of the development project. As a further complication, the loan may be just 'senior' debt or it might include a layer of 'mezzanine' finance. The 'lead' bank, usually an established property lending bank with the necessary expertise employed in-house will arrange the syndication of banks. The 'lead' bank may underwrite the entire facility or agree to use its 'best endeavours' to secure the syndication. It is usual for the 'lead bank', who may partici- pate in the syndicate, to have the final responsibility for making decisions on behalf of the syndicate during the period of the loan.

(iv) Interest rate options   The development company may wish to protect themselves from the risk of changes in the base rate or LIBOR over the period of the development project, particularly when the market is uncertain. In this instance, the developer may seek fixed-interest loans, but it must be remembered that the development company may be tied to a high rate of interest for a long period, so that they may be unable to gain from subsequent reductions. Therefore, developers may compromise and try to 'hedge' the risk that interest rates will rise during a develop- ment, but this will always be at a cost. The usual form of interest rate hedging is the 'cap', which limits the amount of interest the developer will have to pay and is similar to an insurance policy. It is an interest rate 'option' which has to be paid for at the outset, either to the bank provid- ing the loan or another bank altogether. For instance, with a loan linked to LIBOR, the bank will reimburse the developer the cost of interest over and above the 'cap' rate. Hedging is more difficult on a speculative development loan than on loans for income-producing investment prop- erties due to the uncertainty of the amount of loan outstanding at any one time.

4.3.3   Mortgages

As we have already discussed at the beginning of this chapter, mortgages provided the most usual form of long-term development finance until the 1960s. A mortgage is a loan secured on a property whereby the borrower has to repay the capital loan plus interest by a certain date. However, not many lenders are interested in long-term non-equity participating loans such as mortgages. From the lender's point of view a mortgage is a fixed income investment and very illiquid. Some banks, the larger building societies and many life or insurance companies provide mortgages on commercial properties. However, the availability of mortgages is limited due to the problem of the 'reverse yield' gap. It is rare to get fixed rate interest mortgages, although some life funds provide long-term fixed-term interest mortgages depending on prevailing interest rates. Mortgages regained some popularity in the late 1980s as long-term interest rates were low but by the early 1990s long-term rates rose again.

Mortgages may normally be granted on a loan to value basis of between 60 and 80% depending on the risk involved. The amount of mortgage secured will depend on the security being offered by the borrower in relation to the quality of property, the financial standing of the tenant and the borrower. Mortgage loans are normally 20–25 years in line with the length of occupational leases.

Various methods have been developed to overcome the initial 'deficit' problem caused by the difference between rental income and interest repayments over the first 5 or 10 years. Interest payments may be fixed for a certain period and then converted into a variable rate. Some borrowers do not want to be exposed to variable interest rates and may negotiate what are termed 'drop lock loans' which allow the borrower to switch from a variable rate of interest once the rate reaches a certain level.

4.3.4   Corporate finance

As we have already discussed (Section 4.2.4), there are various methods available of raising equity and debt finance from institutional investors via the Stock Exchange, which we will now examine briefly.

(a) Equity finance

(i) **New shares**   Companies may raise money by selling shares to investors in a floatation on the Stock Market or the Unlisted Securities Market. The majority of new share issues are underwritten by financial institutions for a fee, who will buy any shares not bought. Several new property companies became quoted during the favourable Stock Market conditions of 1993 and the first half of 1994.

There are two types of shares: ordinary and preference shares. An ordinary share is a share in the equity (ownership of the company). Ordinary shareholders have voting rights and share in the risks and profits of the company. Profits after tax are distributed via dividends, usually half yearly. Companies may also issue convertible preference shares at a fixed dividend which, within a specific period, may be converted into ordinary shares. Preference shareholders rank above ordinary shareholders in entitlement to dividend payments. However, preference shareholders do not participate in any growth in the company profits and normally have no voting rights.

(ii) Rights Issues   A company can raise additional capital by offering existing shareholders the right to purchase a number of additional shares in proportion to their existing shareholding at a lower price. As with new issues a rights issue is normally underwritten. The net asset value (NAV) per share will be diluted. The ability of a company to raise capital via a rights issue will depend on Stock Market conditions, the state of the property market as measured via property share performance and the NAV per share. During 1993 and the early part of 1994 several property companies successfully raised capital on the Stock Market via rights issues as property share prices performed well, while only marginally diluting the NAV of their shares.

(iii) Retained earnings   One source of finance is the company's own resources generated by profits.

However, some profit will need to be distributed to shareholders as dividends. How much of the profits is paid out as dividends or retained is up to the company to decide, but they must be aware of the interest of their shareholders in maintaining a reasonable dividend.

(b) Debt finance

Debt finance instruments may be secured on specific property assets or the property assets of the company as a whole. Alternatively, they may be unsecured where investors have to rely on the financial strength and track record of the company.

(i) Bonds   A bond is effectively an 'I owe you' note, secured on a specific investment property or completed and let development owned by the company. Investors in a bond receive interest each year and their initial investment is repaid at a specific date in the future. Bonds are securities which can be traded on the Stock Market. The interest payments (known as the coupon) can be structured to avoid the usual problem of rental income shortfall. With 'stepped interest' bonds the investor receives a low interest rate initially which rises at each rent review. An alternative is a 'zero coupon' bond where no annual interest is repaid but the investors are repaid on the redemption date at a premium. However, both these types of bonds rely on rising property values and many such bonds issued in the late 1980s are having to be restructured through negotiations with the investors.

(ii) Debentures   Debentures are securities which can be traded on the Stock Market. Debentures are issued by companies to institutional investors whereby the institution effectively lends money at a rate of interest below market levels in return for a share in the company's potential growth. The money is typically lent long-term, usually up to 30 years, at a fixed rate of interest and is secured upon the company's property assets. Normally, the security is specifically related to named properties, but sometimes provision is made to allow the company to substitute one property for another subject to agreement on valuation.

(iii) Unsecured loan stock   Property companies may issue to institutions unsecured loan stock (not secured on the assets of the company) at fixed rates of interest which, within a specific period, can be converted at the option of the institution into the ordinary shares of the company.

4.3.5   Unitization and securitization

The property markets and financial markets are continually trying to develop equity financing techniques to reduce the problem of the illiquidity of property investment. They are attempting to make property investment more comparable to other investments. The late 1980s saw a number of attempts to increase the liquidity of property investments as the size of development projects grew. As we have already discussed, financing large, single developments worth over £50 million with funding institutions is difficult for developers unless two or more funds become involved. This means that often larger single properties are valued at a discount compared with smaller investments.

Many sources within the industry see securitization and unitization of property investment as a way to provide a wider market for property beyond the existing financial institutions who are large enough to participate. Unitization in this context means the splitting up of ownership of a property or a portfolio of properties amongst several investors. Securitization is a general term used to describe the creation of securities which can be traded on the Stock Market, e.g. shares, bonds, debentures and unit trusts. Therefore the creation of securities is one way of achieving unitization. We shall now examine the various securitization and unitization techniques which have been introduced so far.

The Landsdowne Building, Croydon - 110 000 sq ft (10 219 sq m) headquarters office building by Lynton plc and completed in 1991. The entire building is let to Tarmac and won the Croydon Design Award for its architecture.

There have been various attempts at 'unitization' of large properties, i.e. splitting the ownership of the property into small manageable chunks, allowing several investors to invest in the property. However, it has proved very difficult in practice to divide ownership of an individual property in law.

Attempts at the unitization of individual properties have included SPOTS (Single Property Ownership Trusts) and PINCS (Property Income Certificates). SPOTS involve a trust owning a property and spreading the ownership among investors in the form of units similar to unit trusts; however, this method faltered due to tax problems. The Inland Revenue could not be persuaded to allow the income from the property to pass to the unit holders without prior deduction of tax. PINCS involve a complex structure of companies and leases. A PINC is a security consisting of an income certificate, a contract to receive a share of income of the property after management costs and tax, and an ordinary share in the management company, which manages the property. As a security, it is capable of being traded on the Stock Exchange. The investors receive the benefits of ownership, in the form of a share in the income and capital growth, without owning the property direct. A PINC is not subject to income tax or capital gains tax. Although PINCS were ready for launch in 1986, they were abandoned in 1989 due to stock market conditions. It is widely believed that once conditions are right, PINCS will be launched, but there is much debate about their likely future success as doubts are raised about the attractiveness of PINCS to investors. PINCS would be available to the general public if adequate regulatory protection measures were introduced by the government.

In 1988 an attempt was made at unitization via a company structure, where a company invests in a single property. In this case, Billingsgate City Securities invested in a single property, Midland Montagu House, an office building in the City of London. It involved three different classes of security: deep discounted bonds, preference shares and ordinary shares. The preference shares were available to the public but did not prove popular. The issue was not successful and, as a result, the owner had to take over the majority of the issue. Single asset property companies (SAPCO) have an ordinary corporate structure and, therefore, have tax problems as the company attracts income and capital gains tax.

There are ways of unitizing a portfolio of properties via unit trusts aimed at small pension funds and private investors. There are two main types of unit trust: authorized property unit trusts (PUTS) and unauthorized unit trusts. PUTS invest directly in property on behalf of their investors which may include private investors. They are strictly regulated to ensure that they invest in a diversified portfolio of low risk prime income producing property as part of an overall balanced portfolio including property securities. A PUT is treated as a company for the purposes of corporation and income tax, but is exempt from capital gains tax. Very few have been created so far. Unauthorized unit trusts are unregulated trusts investing in a mixed or specialized portfolio of properties. They are attractive to tax exempt financial institutions such as small pension funds and charities who lack the funds to invest in property directly. Where all the investors in the trust are tax exempt then the trust is exempt from capital gains tax. Unit trusts are managed by a committee of trustees elected by the unitholders (investors) under the terms of the trust deed.

There is much debate in the industry about the role of securitization and unitization in providing long-term finance for property. A widely acceptable technique has not emerged but PINCS has come the closest to overcoming the problems associated with property investment yet maintaining as many of the benefits as possible of holding property direct. It will take considerable time for an acceptable technique to be introduced where investors tend to be conservative. They need to be convinced of the marketability and performance of such methods of investment.

4.4   THE FUTURE

At the time of writing (December 1994) the development industry is going through a process of restructuring and recovery following the crash of the early 1990s, with many companies reducing their gearing by selling assets or raising capital through new share issues. Many of the financially strong companies and some new players are embarking upon development schemes on the basis of pre-lets and, in some cases, speculative schemes, and financial institutions are providing some of the necessary development finance. However, this must be set in the context of a market which is still over supplied with secondary space and experiencing low rental growth (with some exceptions were there is a marked shortage of prime space). Yields have been driven down in recent years by the sheer weight of institutional money entering the market on the expectation of recovery. Yields are now remaining level awaiting occupier demand and rental growth which is taking longer than expected to filter through.

The main issue is who is going to buy the completed developments of the late 1980s, which are over-rented or still vacant, and remain a burden to both their developers and the banks who financed their construction? The banks have been willing to restructure loans in the interim taking the longer-term view. However, there is much debate over how the refinancing will take place. Although property investment is back in favour with some of the institutions they are not going to be able to provide all the equity finance required to pay off the debts. The stock market has relieved some of the burden through rights issues but this is slowing down. There is a limit to the amount of debt to equity swaps that can take place as the banks dislike holding equity. Banks will remain as short- to medium-term debt financiers so there is a need to widen the range of equity (long-term) sources of finance. Securitization and unitization may be the answer but it is going to take time and a return of full confidence in both the property and equity markets at the same time. There is much debate about whether development loans can be securitized into debentures and bonds or converted into mortgages but activity in this direction has been limited. A lot is going to depend on the strength of the economic recovery underway and any change in government.

Executive Summary

There are a variety of sources and methods of financing property development both in the short-term or long-term. The choice and availability of funding will depend on the nature of the scheme, how much risk the developer wishes to share, and the confidence of both financial institutions and the banks in relation to the underlying economic conditions at any particular time. The most secure route from the developer's point of view, provided the development is 'prime', is forward-funding with an financial institution, a method which combines both short-term and long-term funding. However, if the developer wishes to retain flexibility, either wishing to retain the investment or sell it when market conditions are favourable, then debt finance is more appropriate in the short-term. The terms and method of debt financing will depend on the financial strength of the developer and the value of the security being offered. A pre-let scheme being carried out by a financially strong developer represents the best proposition. The greater the risk the less likely the developer will be able to obtain debt finance on favourable terms, unless the developer either contributes its own capital or shares the eventual profits. If debt finance is used then both the developer and financier must have regard to the availability of long-term finance and the requirements of property investors. Property has to compete with other forms of investment which offer more liquidity to the investor, so funding and valuation techniques have to be developed to improve the attractiveness of property as an investment.

REFERENCES

Barter, S. (1988) *Real Estate Finance*, Butterworth, London.

Brett, M. (1990) *Property and Money,* Estates Gazette, London.

Brett, M. (1991) *How to Read the Financial Pages*, 3rd edn, Hutchinson Business, London.

Darlow, C. (1989) Property development and funding, in *Land and Property Development: New Directions* (ed. R. Grover), E & FN Spon, London, pp. 69–80.

DTZ Debenham Thorpe (1993 and 1994) *Money into Property*, DTZ Debenham Thorpe Ltd, London.

Gooby, A.R. (1992) *Bricks and Mortals,* Century Business, London.

McIntosh, A. (1994) Inflation: the property market myth, Richard Ellis Research, *Estates Gazette,* **9428**, 123–125.

Rodney, W. and Rydin, Y. (1989) Trends towards unitisation and securitisation, in *Land and Property Development: New Directions* (ed. R. Grover), E & FN Spon, London, pp. 81–94.

Royal Institution of Chartered Surveyors (1994) *The President's Working Party on Commercial Property Valuations*, Royal Institution of Chartered Surveyors, London.

CHAPTER 4 CASE STUDY – FORWARD-FUNDING OF A SPECULATIVE DEVELOPMENT

This case study will provide an example of how a speculative development scheme may be forward-funded with a financial institution (fund). This particular funding arrangement between the developer and the fund is on a 'profit erosion' basis which we have briefly referred to already in this chapter. We will examine the structure of the financial agreement and the extent to which the fund controls both the specification of the building and the lease terms.

SITE ACQUISITION

A joint development company appointed an agent in 1992–93 to identify office development sites in good locations. The brief was to find and acquire sites which were capable of being forward-funded with a financial institution and where there was a clear lack of supply of prime space either on the market and/or in the pipeline. Six sites were purchased between the end of 1993 and 1994, one of which is the subject of this case study. The rationale behind these acquisitions was to secure sites, which were not on the market, at low values ahead of the next development cycle.

The case study site is in an established office location within the Thames Valley to the west of London, close to one of Britain's major motorways and with good rail connections to London. The site is close to green belt land and good quality 'executive housing'. It is near a large business park and several other office schemes developed in the 1980s, all occupied by major plc companies. Previous research carried by the agent combined with a good knowledge of the area revealed that many senior executives of companies lived in the area. Also the local planning authority has a very strict policy in relation to office development which limits supply. However, there is still an over supply of secondary space on the market.

A residential developer owned the site, having only recently outbid the joint venture company in acquiring it from another developer on the basis of obtaining a residential consent. The site had the benefit of an existing planning consent granted back in April 1990 for an office development of 37 000 sq ft (3437 sq m) gross, with 172 car spaces. After months of negotiations the residential developer was persuaded to sell the site as it was worth marginally more for office development than for residential. This was due to the fact that the residential developer was having difficulty in obtaining a valuable enough residential consent in terms of the density permissible.

ARRANGEMENT OF FUNDING

So at the end of 1993 the site was acquired for £2 million. The developer then appointed the agent to secure forward-funding of the speculative scheme on the basis of a 'profit erosion' deal with no bank guarantees. The agent had secured three similar deals at the beginning of 1994 so it was felt that market conditions were favourable to approach some funds. Some financial institutions at the beginning of 1994 were beginning to look at funding well-located speculative 'prime' schemes in specific areas where their research had shown there was a shortage of prime space and good prospects for rental growth. Development funding was being seen as the only way of obtaining prime stock in locations where most new buildings are over-rented.

A package was put together and it did not take very long to secure a fund through the agent's contacts. This particular fund had recently lost out to another fund on another office scheme, which the agent had been appointed on. The fund agreed to fund the scheme on their first site visit. The presentation made to the fund prior to their site visit included an extensive analysis of the local market and that of the Thames Valley in general. Research carried out by the agent revealed that there was only 161 000 sq ft (14 957 sq m) of office development (including this proposal) in the development pipeline: in other words sites with planning consent or schemes being constructed for office development. Evidence of letting transactions in recent months revealed (since August 1993) that rents were beginning to rise in the Thames Valley area in general and were nearing £20 p.s.f. in some locations. At the height of the boom the town in which the site is located had seen rental levels reach nearly £30 p.s.f. (£322.92 p.s.m.) and many major companies had been attracted from London. The results of the research also showed that only 48 000 sq ft (4459 sq m) of prime office space was available on the market. Evidence was also presented of a number of active requirements for offices in the area.

The package also included the results of a site investigation and environmental audit which showed no evidence of the existence of contamination. As we have already discussed in Chapter 2, such environmental audits are a prerequisite to the arrangement of development finance.

PLANNING

The developer's proposal to construct 37 000 sq ft (3437 sq m) gross of offices in two buildings was similar in character and size to the existing consented scheme for a single building. Accordingly, planning permission was granted very quickly by the local planning authority. A full planning application was made in March 1994 and consent was granted only 2 months later.

TERMS OF FUNDING AGREEMENT

Heads of terms in relation to the funding agreement were agreed between the developer and the fund in April 1994. They were as follows.

(i) A base rent of £18 p.s.f. (£193.75 p.s.m.) was agreed on the basis of air-conditioned buildings. This was slightly above the level most recently achieved on lettings in the town but the fund were convinced by the agent's report on lettings in the Thames Valley area and that such evidence pointed to a rental level nearing £20 p.s.f. (£215.28 p.s.m.) very soon. If the rent achieved is above £18 (£193.75) and below £22 p.s.f. (£236.81 p.s.m.) then the slice of rent over the base rent (the 'overage') will be capitalized at the agreed yield and shared equally between the two parties. If the rent achieved is between £22 (£236.81) and £24 p.s.f. (£258.34) then the developer only receives 15% of the overage capitalized at the agreed yield. This allows the developer and the fund to share any rental growth over the period of the development and by not capitalizing any rent over £24 p.s.f. (£258.81 p.s.m.) the fund ensures the developer is not motivated to over-rent the investment.

(ii) A yield of 7.25% was agreed based on evidence of comparable investment transactions on completed buildings at 6.25%. The 1% discount represents the risk the fund is taking on in relation to a speculative development funding.

(iii) All the interim or short-term finance would be provided by the fund at an interest rate of 7%, compounded quarterly in arrears. Interest would be rolled up on all moneys advanced by the fund until the earlier of 3 years from the date of completion of the scheme or the letting of both buildings, with due allowance for any rental income.

(iii) The developer and the fund agreed a schedule of development costs to include site costs, building costs, professional fees and all other development costs (including void costs) together with the interest. The schedule showed a total development cost of £6 million which is referred to in the agreement as the 'Maximum Advance'. This represents the maximum amount of money the fund is prepared to advance to finance the development. If the development costs exceed this sum then the developer is responsible for providing the shortfall. In addition each element of the total development cost is capped, which is quite a penalty for the developer. Even if the developer saves money, e.g. on the building cost element, this saving cannot be credited against a cost over-run elsewhere. In addition, the developer is under an obligation to ensure that the value of the scheme does not fall below the 'Maximum Advance'. This means that if the rental achieved falls below approximately £15 p.s.f. then the developer has to compensate the fund for the loss of value.

(v) The site value was agreed at £2 050 000 which the fund would advance to the developer on the signing of the funding agreement. Accordingly the developer makes an initial profit on the site value of £50 000.

(vi) The schedule of agreed development costs includes a developer's co-ordinating fee of £90 000 (3% of the building cost) to allow for the developer's in-house costs such as project management fees etc. The fund's monitoring costs to cover their costs of monitoring the scheme through their surveyor was agreed at £42 500.

(vii) The agreement also sets out the lease terms and tenant covenant acceptable to the fund. The fund will only allow single lettings of each building on 25 year full repairing and insuring leases with 5 yearly upward-only rent reviews. However, if market conditions dictate then a 15 year lease or a 25 year lease with a break at 15 years is acceptable. The agreement allows some flexibility over single floor lettings. Inducements to achieve a letting at the base rent are allowed provided that any loss in value is deducted from the developer's profit. Potential tenants have to demonstrate that their profits amount to three times the rent or their net assets are 10 times the rent payable.

(viii) It was agreed that the developer did not have to provide any bank guarantees, as the fund was satisfied with the combined financial strength and expertise of the development partners.

(ix) The profit due to the developer is calculated at the earlier of the full letting of both buildings or 3 years from the date of practical completion. At that time the development costs (subject to the maximum advance provisions) are deducted from the gross development value (developer's share of the rent achieved capitalized at the agreed yield of 7.25%) to calculate the balancing payment due. The calculation allows for any rental income received by the fund in the meantime. Also the calculation of development costs allows for all interest that has been rolled up to that date and all holding costs incurred in the void period, e.g. empty rates.

(x) If a letting has not been achieved by the end of 3 years after completion of the scheme then the developer can walk away from the scheme with nothing leaving the fund with the ownership of the building and the responsibility of letting it. There is provision to extend the 3 year period by a further 12 months if the developer is in negotiations with a tenant or the buildings have been partly let.

(xi) The agreement also includes the usual obligations on the developer to construct the building in accordance with the agreed specification and to provide collateral warranties with the professional team.

It took 4 months to agree and sign the legal documentation which was completed in August 1994.

SPECIFICATION

The specification agreed between the two parties proposes good quality buildings of brick construction with air conditioning and may be typical of institutionally acceptable specifications for the 1990s. The specification reflects a movement away from extravagant over-specified internal finishes so often typical of office developments in the 1980s: the emphasis is on a reasonable quality. The institutional investment market on the whole is still conservative and continues to insist on air conditioning to ensure that the building has wide tenant appeal. However, due to the fact many tenants dislike air conditioning, allowance has been made on this scheme for 50% openable windows and the system has been designed to allow the tenant to shut down the air conditioning on whole or part floors. The design of the air conditioning also allows for total flexibility in the tenant's layout of and use of the building which adds to the cost of the scheme but is essential to appeal to as wide a market as possible.

Attention has also been paid to the energy efficiency of the proposed buildings. A BREEAM assessment (an energy efficiency award system based on points administered by the Building Research Establishment, discussed further in Chapter 6) has been undertaken, currently a requirement of both investors and occupiers. Although this building will have air conditioning, which receives a negative score on the assessment, the other elements assessed bring the building up to a suitable standard for a BREEAM award. The tenant will have the ability to install a remote computerized building management system to control the light, heat, power and air conditioning supplied to all parts of the building from a single computer. This is seen by the developer as an essential selling point enabling the tenant to control running costs.

The fund has also insisted on arranging an option to take out decennial insurance for the benefit of tenants against latent defects arising in the building for a period of 10 years from completion. Under this type of policy the insurer will immediately settle any claim for repairs due to a latent defect, and then pursue the professional team if appropriate, for a premium in this case of £30 000 (1% of the building costs). Under this type of policy, even if it is not taken out later on, the insurer needs to carry out an audit of the proposed building and monitor its construction for a fee in this case of £5000.

CONCLUSION

Work on site started in September 1994 on a 15 month building contract. The building contract was awarded at a cost of approximately £75 p.s.f. (£807.30 p.s.m.), which is well within the budget set by the funding agreement. The letting agents have been approached by a potential tenant 1 month into the contract on the basis of a deal equating to £23 p.s.f. (£247.57 p.s.m.). Appraisals of the scheme show the developer is on target to achieve a 23% profit on cost assuming a 12 month rental void. There is also just over 3 years cover before the profit erodes to nil if a letting is not achieved.

From the fund's point of view this funding arrangement should produce a prime investment in a good location, in accordance with their specification and let to acceptable tenants. The deal is structured to ensure that the fund's exposure to risk is quantifiable and safeguarded, whilst allowing a share in any uplift in rental values. The agreement motivates the developer to build the scheme within budget and on time. The developer will profit provided this aim is achieved and the building is let on satisfactory terms within 3 years. This forward-funded deal reduces the developer's exposure to risk without the need to provide guarantees, but at the price of sharing rental growth over the period of the scheme and selling the scheme at a discounted yield.

5

PLANNING

5.1 INTRODUCTION

The system of statutory planning control in the UK is one of the most sophisticated in the world. It is necessary to obtain a specific planning consent for virtually every development. The only exceptions, while large in number, are of a relatively minor nature normally covered by General Development Orders. It is beyond the scope of this general book on property development to deal in detail with all aspects of town planning. However, as the obtaining of planning permission is a critical part of the property development process this chapter will provide at least a general appreciation of the principles involved. Whilst the planning process determines the value of a development, it adds uncertainty to the development process in the form of frustrating time delays, and cost increases and unpredictable decisions. Developers are concerned about these and other aspects of the planning process. During the 1980s numerous changes were introduced to the planning system by the government to speed up the planning process and encourage enterprise; some of these – by increasing flexibility – also increased market uncertainty.

5.2 PLANNING AND THE ENVIRONMENT

Since the mid-1980s the government has changed the emphasis of its general policies in the light of increasing environmental concern. This has lead to more rigorous study of development proposals, wider involvement of third party consultees and there are often consequent delays in the consideration of the planning application.

The more prominent role of environmental considerations was a consequence of the debates held at the 1992 Earth Summit in Rio de Janeiro and subsequently the government's commitment to the concept of 'sustainable development', i.e. that development which would not result in the best of today's environment being devalued for future generations. In 1988 the European Parliament issued a directive requiring the environmental impact of a large number of developments to be formally assessed and the subsequent publication by the government of the Town and Country Planning (Assessment of Environmental Effects) Regulations set out specific requirements in this regard (see Section 5.6).

In the late 1980s and early 1990s, therefore, there has been increased linkage between town planning and environmental law and regulations. The key town planning legislation is the Town and Country Planning Act 1990 as amended by the 1991 Planning and Compensation Act, and the Planning (Listed Buildings and Conservation Areas) Act 1991. The details of the legislation and its interpretation are covered in more specialist books which you will find listed in the bibliography. However, this chapter represents a broad summary of how the planning system operates and its implications for those involved in the property development process.

5.3   THE PLANNING SYSTEM

The planning system is overseen in the government by the Secretary of State for the Environment (and in Wales by the Secretary of State for Wales) who, in the context of the legislative framework provided by the Planning Acts, issues statements of policy and guidance (through ministerial Circulars and Planning Policy Guidance Notes) to local planning authorities on relevant planning matters, including their approach to applications of varying types. The Secretary of State also has the power to consider matters of strategic policy in county structure plans and Unitary Development Plans (UDPs), although this power is rarely used, and through his own appointed inspectorate considers all planning appeals against decisions of planning authorities. The Secretary of State also has the power to determine any application or planning appeal if it is of national or regional significance via what is known as the 'call in' procedure.

The list of Planning Policy Guidance Notes (including those issued for Regional Planning Guidance) issued by the Department of the Environment is attached as Appendix B. These statements of policy are rigorously scrutinized by those involved in the property development and planning process as they provide general guidance to all concerned (including local authorities) on how to approach planning applications for various forms of development ranging from housing, shopping and business parks to sports facilities and power stations.

In addition to the formal policy statements, the content of speeches and statements from relevant ministers and particularly the Secretary of State for the Environment are important considerations. An example of the importance of such statements is the controversy in 1994 prompted by a speech by the Secretary of State relating to out-of-town retailing. As the issues which arise in the field of town planning (often brought to a head by applications for development) have become more politically sensitive, the role of politicians at all levels of government in the making of planning decisions cannot be over estimated.

The day-to-day operation in England and Wales of the planning system outside London and the former Metropolitan Counties relies on the two-tier system of local government (which is under review) with general strategic policies being set out in the county structure plans by the county councils and the vast majority of planning applications are determined by the district councils in the framework of structure plan policies, the policies of a local plan which the district councils may have prepared and the Secretary of State's general planning guidance referred to above.

In Greater London and the former Metropolitan counties (abolished by the 1985 Local Government Act) the functions of both county and district councils are carried out by single unitary authorities, i.e. those former Metropolitan districts in areas such as Greater London, Greater Manchester, West and South Yorkshire, Merseyside, etc. The town planning role of these authorities includes the preparation of a UDP which combines the strategic nature of a structure plan and the more site-specific nature of a local plan. The plans are prepared against the background of regional planning guidance from the Secretary of State for the Environment. The unitary authorities themselves determine all planning applications in the context of the UDP and the Secretary of State's general planning guidance.

Whilst this chapter will concentrate on the system as applies generally in England and Wales (see comments on Scotland below), there are certain areas which are exceptional:

1. Enterprise Zone Authorities and UDCs. These principally are authority areas designated under the 1980 Local Government Planning and Land Act for primarily reasons of economic regeneration (e.g. see Chapter 2). These authorities perform the functions of planning authorities in the area of designation over the period of their life. Enterprise Zones are normally limited to 10 years whilst the UDCs are anticipated to be wound up by 1998.

2. Simplified Planning Zones. Simplified Planning Zones are particular zones within a planning authority's area where certain planning restrictions may be relaxed (see Chapter 2).

Since 1992 the Local Government Commission has been considering on a county by county basis whether the two-tier county/district relationship is appropriate. Various recommendations have been made, primarily proposing the change of certain district councils to unitary authorities possessing more responsibility for planning and other matters at the expense of the county councils.

The combination of the structure plan and local plan outside London and the former Metropolitan Counties and the UDP in the unitary areas is summarized by the term 'the development plan'. This description may also extend to more specialist plans covering Minerals and Waste Disposal. The importance of the development plan has been elevated by the 1990 legislation not only as authorities are required to produce such plans (prior to the 1990 Act their preparation was discretionary) but because of the increased prominence given to the policies of fully approved or adopted plans in the determination of applications. The key section of the 1990 Act is Section 54A which states:

> In making any determination (of a planning application) under the Planning Acts, regard is to be had to the development plan, the determination shall be made in accordance with the plan unless material considerations indicate otherwise.

In these terms, the role of the development plan is a key one in the planning process. So far as those involved in property development are concerned, much more attention is now placed on the plans prepared by the various authorities. This manifests itself not just in relation to studying the relevant policies within the plans but more and more in the involvement in the preparation of the plans themselves through the making of representations at the appropriate time in an effort to influence their content (see below).

In the development plan process, the county council's most important role is that of strategic policy maker in the structure plan, a document that sets out general policies relating to a county as a whole, normally over a 10–15 year period. This would include the number of dwellings to be constructed in each district in the county, the amount of land required for employment related development, the amount of new retail floorspace required, those areas which should be specified for environmental protection and what new roads are proposed. The county council's proposed policies within a structure plan are the subject of several stages of public consultation (see Figure 5.1) and a public inquiry (the Examination in Public (EIP)) is held where invited interested parties can make representations seeking to change the county council's intended strategy. The EIP is presided over by a 'panel' appointed by the Secretary of State who considers all the arguments involving the county council concerned, the district councils within the county, developers, environmental groups and other local organizations.

Prior to the 1990 legislation, the panel reported to the Secretary of State with their recommendations; the Secretary of State considered the report and then (after a period of consultation) approved the structure plan with whatever changes he saw fit to make. However, that system has been altered in that the panel now report with recommendations to the county council who decide whether to amend the plan accordingly. Whilst the Secretary of State retains powers to intervene, in practice these are exercised only exceptionally. Developers are often critical of this

1. Initial consideration of policy issues, statistics (e.g. household formation) and research by county council
2. Publication of initial consultation draft structure plan
3. Period for public comment and representations on consultation draft plan
4. County council considerations of representations and amendments to consultation draft
5. Publication of submitted draft structure plan for future representations and to the Secretary of State for the Environment
6. Period for public consultations and representations
7. Selected parties invited to attend EIP into the structure plan (and to expand on representations made in representations at stage 6)
8. EIP held with panel appointed by Secretary of State presiding; discussion of issues identified by the panel as being of key significance
9. Panel discussion at EIP of representations made at stage 6; production of report of recommendations to county council relating to changes in policy deemed appropriate
10. County council considers changes to submitted structure plan resulting from panel's report and publish proposed modifications to submitted plan
11. Further opportunity for representations to be made
12. County council considers representations and then approves the plan

Note: Secretary of State may decide at certain stages to formally consider and approve the plan

Figure 5.1   Structure plan approval process.

process as they see the county council as being in a position to ignore any panel recommendations they do not like. Those who favour the system argue that it allows for important strategic decisions to be made at county rather than national level.

Once the structure plan is approved, the county council's role is to ensure that the plan is regularly monitored and reviewed, where necessary. In terms of the submission of planning applications on a day-to-day basis, the county council mainly acts as a consultee to the district councils, with the exception of specific types of planning applications which it has the power to determine (e.g. those dealing with mineral extraction). However, it does retain advisory powers relating to highway matters, which for many major schemes requiring planning permission are vitally important. The district council's role is to consider and determine the vast majority of all planning applications submitted. They are also required to prepare, in the context of the county structure plan, local plans covering their district, where specific proposals are made and policies are set out for the control of development. An example of the relationship between a structure plan and a local plan is where a structure plan requires 5000 dwellings to be constructed in a particular district over a 10-year period to meet strategic needs and the local plan allocates specific sites for development in the district to enable the requirement to be met.

The 1991 Act requires district councils and unitary authorities to prepare local plans or UDPs. The preparation of a local plan follows a series of well-defined stages which allows for public consultation on emerging policies. These are set out in Figure 5.2. The development industry regularly involves itself in each stage and is usually represented when the objections to the plans made during their deposit stage (i.e. the deposit draft plan) are considered at a public inquiry.

The public inquiry is presided over by a Department of the Environment inspector who listens to evidence supporting objection to the deposit draft of the plan and the planning authority's reasons for maintaining the policy of the deposit version of the plan. The Inspector does not, however, have the power to impose his view on the planning authority. A report is compiled by the inspector which makes recommendations to the authority on whether each objection should result in the deposit draft plan being changed. The planning authority then decides whether to act upon the recommendations or not. If they do not then they have to justify their decision but they are not obliged to follow the inspector's views. After making modifications to the plan considered appropriate and a further consultation period the local or UDP can be

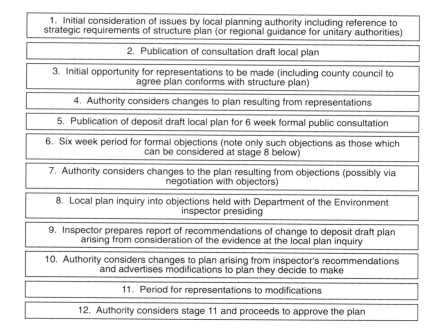

1. Initial consideration of issues by local planning authority including reference to strategic requirements of structure plan (or regional guidance for unitary authorities)

2. Publication of consultation draft local plan

3. Initial opportunity for representations to be made (including county council to agree plan conforms with structure plan)

4. Authority considers changes to plan resulting from representations

5. Publication of deposit draft local plan for 6 week formal public consultation

6. Six week period for formal objections (note only such objections as those which can be considered at stage 8 below)

7. Authority considers changes to the plan resulting from objections (possibly via negotiation with objectors)

8. Local plan inquiry into objections held with Department of the Environment inspector presiding

9. Inspector prepares report of recommendations of change to deposit draft plan arising from consideration of the evidence at the local plan inquiry

10. Authority considers changes to plan arising from inspector's recommendations and advertises modifications to plan they decide to make

11. Period for representations to modifications

12. Authority considers stage 11 and proceeds to approve the plan

Figure 5.2   Local plan/unitary development plan adoption process.

'adopted' thus giving it the prominence referred to in Section 54A of the 1990 Act.

This lack of obligation to follow Inspectors' recommendations is a source of frustration to developers, particularly if they have been successful in persuading the Inspector to recommend that changes are made to a deposit draft plan after hearing evidence at the public inquiry. The contrary view is that it is right for decisions of this nature to be made by local/unitary authorities on the basis that local democracy is seen to take precedence on planning matters.

As set out in Figures 5.1 and 5.2, whilst there are ample opportunities for involvement in the evolution of structure plans, local plans and UDPs, the time taken for the processes to run their course is considerable. It is not unusual for the entire structure plan process to take 2 years to complete from the initial consultation to the final approval whilst a local plan or UDP can often take over 3 years. A typical example is the Wakefield UDP which had the following timetable:

Publication of first consultation draft for period of presentation –	February 1991
Consideration of representations –	April–October 1991.
Publication of deposit draft plan –	November 1991.
Public inquiry into objections received to deposit plan –	October 1992–January 1993.
Receipt of inspector's report –	February 1994.
Period of modifications to the plan following Inspector's report and additional consultation –	July–September 1994
Final adoption of plan –	December 1994.
Total time elapsed –	3.75 years

The delay, uncertainty and cost which is an inevitable consequence of the process is a constant problem to developers, particularly as it is common-place for there to be slippage in the local authority's envisaged timetable. The government has made it clear in its Planning Policy Guidance that the determination of planning applications must not be delayed pending the approval or adoption of development plans and that only if a proposal is so significant that it may prejudice the decision-making process should it be regarded as 'premature' to the finalization of the plan. Similarly, government policy allows for account to be taken of plans as they progress through the various stages with more weight being attached to their policies the further along the process they proceed.

However, particularly in the areas of the out-of-town retail, business parks and residential development where 'green field' sites are proposed for release by developers it is often the case that such proposals are having to be pursued through the development plan process rather than through immediate planning applications. To that extent the timing of preparation of a development plan has become, since the 1990 Act, a considerable factor in the site acquisition process for developers.

The operation of the planning system in Scotland is similar to England and Wales in relation to the submission of planning applications, the preparation of development plans, the powers of planning control and the general appeal process. The differences mainly relate to the system of local government under the jurisdiction of the Scottish Office and the Secretary of State for Scotland. The administrative system will also change in April 1996 as a result of the Local Government (Scotland) Act 1994.

Currently the strategic role (the county council's role in England) is performed by the regional councils, e.g. Grampian, Strathclyde, etc., who cover wider geographical areas than their English counterparts. They prepare structure plans as well as involving themselves in regional initiatives. District councils in Scotland perform the day-to-day administration of planning applications and prepare local plans in conjunction with the structure plans for their regions and Scottish Office Planning Circulars.

In April 1996 the two-tier system (under the Scottish Office) is to be abolished and replaced with a single tier or unitary system. The unitary authorities are generally speaking based on the old 'districts' with certain exceptions, for example the former Highland Region is to be a unitary authority. In terms of planning responsibilities, there is no obligation on the new authorities to prepare unitary plans; instead the Scottish Office has instructed the authorities to have comprehensive local plan coverage for their areas (in one or more plans per authority).

In structure plan terms, the intention is for groups of unitary authorities to co-operate (via joint planning boards) in the preparation of a plan for their area which need not follow the boundaries of the former regions. This has led to some confusion, indeed concern that potential joint planning boards may not have the expertise to produce such plans.

Whilst the day-to-day operation of the system in relation to development control matters is likely to remain unchanged, the development plan function and responsibilities will be substantially altered by the new administrative system.

5.4   WHAT IS DEVELOPMENT?

The statutory definition of development as set out in Section 55 (i) of The Town and Country Planning Act 1990 is:

the carrying out of building, engineering, mining or other operations in, on, over or under land,

or

the making of any material change of use in any buildings or other land.

Thus, in broad terms, development might be thought of in two categories, one being the carrying out of physical operations such as building or engineering works and the second the making of a material change of use. A book concerned with property development naturally tends to concentrate on the physical activities, but the question of use is of vital importance. Indeed it is possible to argue that planning control is basically one of land use, because once the use of land has been determined, the question of precisely what is built is a matter of detail albeit most important.

For purposes of the planning control over changes of use in land and buildings the key document is the Town and Country Planning (Use Classes) Order 1987. The Order divides various land uses into classes, each separate class consisting of a group of very similar uses. Normally a change from one use to another within the same use class will not constitute development, whereas a change from a use which falls within one class to a use in another will (with certain exceptions) constitute development for which planning consent will be necessary. In practice, buildings are often suitable for a variety of different uses and a change from one use to another might involve a change of use for which planning consent will be necessary. The value of such a building might be substantially affected by the possibility of being able to obtain the necessary planning consent to change from one use to another. For example, a warehouse (Class B8) on the edge of town may be suitable for retail uses like DIY (Class A1).

The 1987 Order replaced an earlier 1972 version as a recognition of the inflexibility of the old Order in relation to modern industry and its use of buildings. Appendix A contains a summary of the 1987 Use Classes Order.

One of the most significant changes made was to incorporate offices (other than those in Class A2 providing financial and professional services), research and development and light industrial uses within one class, 'B1 Business', provided that the use can be carried out in any residential area 'without detriment to the amenity of the area by reason of noise, vibration, smell, fumes, smoke, soot, ash, dust or grit'. This change was made in recognition of the needs of modern industry, particularly 'high-tech' industries who to varying degrees often carry out office, research and development and light industrial uses within one building. However, during the 1980s development 'boom' it had a dramatic effect on industrial land values as landowners expected office values for industrial sites. Many local authorities resisted the change but were not supported by the Department of the Environment Inspectorate at planning appeals. The complexity of the uses to which land and buildings can be put linked to changing trends in the types of usage means the effectiveness and relevance of the Use Classes Order is constantly under review.

It should also be pointed out that the Use Classes Order is by no means comprehensive in covering all uses and there are commonly disputes between applicants and local authorities over what are known as 'sui generis', i.e. uses not within any defined class in the Use Classes Order.

On the face of it, the description of the physical works which need planning consent appears straightforward. However, as is the case with virtually all legislation, circumstances will arise in which there is room for argument over the interpretation of the statutory provisions and a considerable volume of case law has been built up as to what constitutes building or engineering, mining or other operations for the purpose of planning control. Arguments are often concerned with the placing of mobile structures on land or the installation of plant and machinery in a building or physical works carried out inside a building (which in no way affect the external appearance) which do not need planning consent, assuming that in both cases no material change of use results or the building is a listed building.

5.5   THE PLANNING APPLICATION

Anyone wishing to carry out a development for which planning consent is required must apply to the local planning authority normally (either the district council or unitary authority) for the necessary consent. At the outset, it is prudent to check that a planning consent is necessary. A development will not require planning permission if it falls within the provisions of the General Development Orders which are regularly extended so as to remove a wider range of developments from planning control and the necessity for planning consent.

In some cases, where it is not immediately clear if a planning consent is required, the planning authority can be approached for informal advice. It is also possible to formally establish whether a proposed use of land or buildings or operation is lawful by applying to the local planning authority for a formal 'certificate of lawfulness', as to whether the proposals would constitute development for which a planning consent is required under Section 192 of the Town and Country Planning Act 1990. The local planning authority must give a formal decision within 8 weeks and there is a right of appeal against the decision.

Anyone may apply for planning consent in respect of any property. It is not necessary for the applicant to have any legal or financial interest. However, if the applicant is not the owner of the property, a notice must be served on the freeholder, on any lessee with an unexpired term of at least 7 years still to run and on any occupier of agricultural property, advising them of the application for planning consent. In practice, the planning authority has a set of forms upon which these notices must be served, and the applicant must advise the planning authority of the names and addresses of the people on whom they have been served. In certain instances of development, e.g. buildings for various types of public entertainment or buildings exceeding certain heights, these are publicized by the local planning authority putting a notice in the local press and the notice makes it clear where any member of the public might inspect the plans showing the proposed development. Where the development might have a significant effect on neighbouring property, the authority is obliged to draw the attention of neighbours to the application. Any owner or lessee or occupier of agricultural land upon whom notice is served, or any other member of the general public, has the right to make representations to the local authority.

There are several types of application which can be submitted to a local planning authority dependent upon the type of development which is proposed and the location of the site concerned. These principle types of application are as follows:

1. Outline planning application. An outline application seeks to establish the principle of a particular form of development outside Conservation Areas, without the need to deal with the matters of the siting of the buildings, their design, external appearance, landscaping or the means of access into a site. These 'reserved matters' can either in whole or in part be left for the future submission and determination of the planning authority in the event of outline permission being granted. A typical example of such an application would be for the residential development of a greenfield site on the edge of a settlement.

2. Full or detailed application. A detailed application will seek not only to establish a land use principle but approval for all the reserved matters listed above. The application itself is thus comprehensive and in the case of a residential development would include not only the information about the location of the site but also the layout of houses and roads, the design of the dwellings themselves, the principle landscaping proposals and all the relevant technical information. As the design of new buildings and other reserved matters are particularly important considerations in Conservation Areas (see below) it is mandatory to submit full applications in such areas.

3. Changes of use. Legally an application for the change of use of land or, more commonly, buildings is regarded as a full application rather than an outline. This requires an applicant to submit full details of

their proposals to the planning authority. Changes of use applications are normally (though not always) related to the uses defined in the Use Classes Order where such permission is required to change from one category of use to another. A typical example would be the change of use from a Class A1 retail shop on a high street to a Class A2 professional/financial use such as a building society.

4. Applications for listed building consent or Conservation Area consent to demolish. All proposals to demolish buildings in Conservation Areas requires specific permission and thus proposals for development in a Conservation Area must also, if demolition is envisaged, be accompanied by a separate application for Conservation Area consent to demolish. Similarly any changes proposed to listed buildings, i.e. those specifically identified by the planning authority and English Heritage as being worthy of protection, must be the subject of specific applications.

A developer wishing to carry out building or engineering works should consider the advisability of applying for an 'outline' consent which establishes the principle of development proposed. The decision whether to submit an outline or detailed application will depend upon the location of the site, the issues involved, and the nature of the developer's legal interest in the site. For example, the developer may only have an option to acquire the freehold of the site or a conditional contract which is triggered by the granting of a satisfactory planning consent. Preparation of the necessary design drawings for a large building project can be costly and the submission of an outline application may in the end save a good deal of time and trouble. An outline application must give sufficient information to describe adequately the type, size and form of development proposed. The local planning authority will reserve for subsequent approval the reserved matters and will attach conditions to an outline planning consent to cover such issues. A typical condition (which will be one of many on an outline permission) would be as follows:

Before any development is commenced, detailed plans, drawings and particulars of the layout, siting, design and external appearance of the proposed development and means of access thereto together with landscaping and screen walls and fences shall be submitted to and approved by the local planning authority and the development shall be carried out in accordance therewith.

Fees are payable to the local planning authority in respect of applications and the appropriate fee calculated in accordance with scales prescribed by the Secretary of State under his powers derived from Section 303 of the Town and Country Planning Act 1990 (usually reviewed annually) must be paid at the time the application is submitted.

The fees are charged on a sliding scale depending on the size and nature of the development. Currently, the cost to the planning authorities in processing applications is not met in full by the fees charged.

As described above the planning authority is required to determine each planning application within an 8 week statutory period which runs from the date the application is deposited, together with the correct fee. If no fee or an incorrect fee is deposited with the planning application, the 8 week period will not start to run until the correct fee is paid.

Planning applications must be made on forms provided by the local planning authority. The forms are self-explanatory but many planning applications are delayed because applicants either do not complete the forms accurately or fail to provide all the information which is requested. When the application form is submitted, the planning officer (a qualified person employed by the local planning authority who will handle all matters arising from the application) should be asked to confirm that the form gives all the information required and that no additional information is needed from the applicant and also that no supporting documents are necessary. The application might be one of many received by the planning authority each week and will take its turn in the queue to be processed and checked; the site will be inspected by staff of the planning authority and there is commonly consultation with other organizations, such as the highway authority or the water authority on site restrictions, or with amenity societies in areas of special environmental interest. The planning authority will normally consult other appropriate authorities before giving planning permission. For example, the Highway Authority will be consulted on the design of any new roads which are ultimately intended for adoption as public highways, or the local Environmental Health Officer might be consulted on potential noise or fume problems, or the Health and Safety Executive might be consulted on 'hazardous' or potentially explosive processes that might be proposed in the development.

There are various matters upon which it is important that developers should satisfy themselves by means of direct discussions with the appropriate authorities. The more important of these are the adoption of roads and footpaths by the Highway Authority where it is appropriate for them to be adopted; the acceptance by the local authority of public open space provided within any development; and the necessary approvals under the building regulations, fire regulations or any other statutory provisions which relate to the development in question, e.g. the Factories Acts, the Shop Acts and Public Health Acts. Developers will agree direct with the Highway Authority the design and specification for all new roads and footpaths and enter into the necessary adoption agreement. They will reach a similar agreement with the appropriate water company in respect of sewers. A planning permission in no way constitutes an agreement for adoption.

During this process the staff of the planning authority might wish to discuss with the applicant points that need clarification or problems that are revealed. At this consultation stage, it is often possible to make modest corrections or alterations in the application to avoid such problems, but more radical alterations might require another round of consultations, or a new application. At the time of submitting the application, it is useful to ascertain the date of the council's planning committee at which the matter will be considered and to check at the appropriate time that it has been placed on the committee agenda.

As discussed above the local planning authority pays particular regard to the provisions of any approved development plan as set out in Section 54A of the 1990 Act and any other material considerations in reaching its decision. However, if it is proposed to grant a planning permission which will constitute a substantial departure from the development plan, the local planning authority must advertise the planning application and give opportunity for objections to be made. The Secretary of State must be advised of the proposals and may decide to determine the application, i.e. to arrange a public inquiry to hear the arguments for and against a particular proposal. In the absence of any intervention by the Secretary of State in such cases, the local planning authority may grant permission.

The Secretary of State has the absolute discretion to 'call in' an application, so that it can be considered by the Department of the Environment. In practice, the Secretary of State intervenes when it is considered that the application is of national or regional significance, for example, shopping developments over 100 000 sq ft in size or proposed new settlements.

5.6   ENVIRONMENTAL IMPACT ASSESSMENT

Since the European Community Directive 85/337 and the Town and Country Planning (Assessment of Environmental Effects) Regulations 1988 introduced the need for environmental impact assessment in certain instances, a developer now may be required to produce an environmental statement as part of a planning application.

This statement should examine the impact of a development proposal on the environment in its widest sense. There are a comprehensive list of factors which can be assessed dependent upon the type of development proposed. Typically this can include analysis of the following aspects:

1. Impact on the landscape
2. Visual impact
3. Ecology
4. Geology
5. Archaeology
6. Air and water pollution
7. Contamination of land
8. Noise pollution
9. Wildlife conservation
10. Agriculture

Some projects, such as oil refineries, motorways and major power stations require assessment in every case. There are about 80 types of developments listed in the regulations which are subject to assessment where they are judged likely to give rise to significant environmental effects, for example, mineral extraction, and major infrastructure projects such as roads and harbours. A developer may apply to the local planning authority for an opinion on the need for environmental impact assessment prior to a formal planning application. A developer has a right of appeal if they disagree with a local authority's request for an environmental statement. The Secretary of State may make a direction that an environmental impact assessment is required or in certain cases the developer may voluntarily prepare an assessment to allay any anticipated fears on environmental matters. The main difference to the normal planning process where an environmental impact assessment is required is that there are additional publicity requirements and an extended period of 16 weeks for the planning authority to determine the application.

As stated in the Introduction to this chapter, the role of environmental considerations in the planning process has increased greatly since 1988. This has taken the form of a general statements in Planning Policy Guidance Notes, particularly PPG1 relating to the need for development to be sustainable in environmental terms and PPG13 which advises that the location of new development should take account of the need to reduce pollution generally, CO_2 emissions in particular and thus travel by road. Whilst such statements from the government are intended as guidance they are open to wide interpretation and developers often find themselves at odds with planning authorities over such matters of interpretation.

5.7   CONSULTATION

It is important to consult the relevant planning officer of the district council or appropriate authority at the outset, before any formal planning application for a reasonably sized development is submitted – often a great deal of wasted time and effort can be saved by so doing. Through early consultations it might be possible to agree the development proposals in principle with the planning officer. They do not always do so, but planning authorities have the power to delegate various powers of decision to their planning officer, thus applicants should always ascertain whether the planning officer is authorized to issue the necessary planning consent or whether a recommendation will be made to the planning committee. The planning authority aim to give a formal decision within 8 weeks of the receipt of the application or such longer period as the applicant may agree. In practice any substantial application takes longer than the statutory period to reach a decision. If consent is then refused or granted subject to conditions to which the applicant has objection, an appeal may be made to the Secretary of State for the Environment. If no decision is given within the prescribed 8 week time limit, the applicant can assume that planning consent has been refused and appeal made to the Secretary of State accordingly. This right of appeal is not universally applied by developers if they consider further negotiations beyond the 8 week period will result in planning permission being obtained.

Even though consultation with the planning officer does not guarantee agreement that a proposal is acceptable, applicants should at least thoroughly understand the views of the planning authority, and their attention will normally be drawn to all the relevant development plans and other policies, so that they are in a better position to consider how their application may be approached.

Some planning authorities appoint advisory committees to comment on certain types of development. Most commonly, architectural advisory panels are set up to give advice on the architectural merits of proposed developments which are to take place in sensitive areas. Planning authorities must also consult local residents' organizations or amenity committees of various kinds. Amenity societies are normally consulted when development is proposed within Conservation Areas and the local planning authority has a duty to advertise the receipt of planning applications for developments in those areas. A local conservation area committee might exist and its comments will carry weight. The planning officer should be able to indicate whether it is possible or advisable to consult such committees or voluntary organizations at the informal pre-application stage.

With environmental issues now high on the political agenda, it is becoming increasingly important for developers to consult with the relevant pressure groups at an early stage. Although planning authorities are advised by the government not to exercise detailed control over design (except in conservation areas), such issues are commonly disputed. This is because matters of design are subjective and capable of being interpreted in several differing ways.

The particular sensitivity which is associated with applications for development within defined Conservation Areas or affecting listed buildings arises because of the special protection afforded to such areas and buildings by the planning legislation. Planning authorities have powers to define Conservation Areas, i.e. areas of special architectural or historic interest which it is desirable to preserve or enhance, and this process can take place in conjunction with the preparation of a local plan or UDP or in its own right. Once defined, all planning applications for development within the area must be detailed and, as set out above, separate applications must be made for any demolition. Government policy in PPG15 is that all proposals for development within the Conservation Area must preserve or enhance its character and there is commonly much debate between planning authorities, conservationists and developers over the design merits of schemes in Conservation Areas in the context of government policy.

If the development proposals entail the demolition or alteration of the character of buildings which are on the list of buildings of special architectural or historic interest prepared by central government, i.e. listed buildings, then special procedures are necessary. As described above, a specific consent for the demolition or alteration of the listed building must be obtained before any work can be carried out. Applications for the necessary consent must be made to the local planning authority and advertised, so that any member of the public may make representations. There are a number of reputable groups with an interest in the protection of listed buildings (who are consulted by planning authorities, e.g. English Heritage) and proposals for alterations or demolitions of listed buildings. Developers may have to spend some considerable time discussing their proposals with interest groups, in an attempt to reconcile their commercial requirements and the sensitive refurbishment of a listed building.

If buildings appear under threat from development proposals, a local authority can issue a temporary listing known as a 'building preservation notice' which has immediate effect and lasts for a period of 6 months to enable the Secretary of State to decide whether to formally list the building concerned. If the building is not subsequently listed there may be compensation payable to the developer for consequential loss.

Local planning authorities also consult archaeology interests, e.g. the Museum of London. In the late 1980s development 'boom' significant archaeological finds were made (e.g. the Rose Theatre in Southwark, London, in 1989). If the site is of archaeological interest, then the developer must allow an archaeological dig or a watching brief by archaeologists to take place before development commences. The period of the dig and any compensation or contribution from the developer is often a matter of negotiation between the developer and the interest group. Under the Ancient Monuments and Archaeological Areas Act 1979, specific areas of archaeological importance can be entered and excavated for up to 6 months but often developers offer contributions to speed up this process. Here again, the development process may be delayed and, therefore, it is prudent for a developer to consult with archaeological groups prior to purchasing a site. Many archaeologists now act as private consultants to developers. The general policy guidelines on archaeological matters are set out in PPG16.

If any trees on site are protected by a Tree Preservation Order, it is necessary to obtain consent before felling or lopping them. Normally the necessary application is made to the local planning authority but in certain cases the consent of the Forestry Commission is necessary.

The Advertisement Control Regulations enable planning authorities to exercise a very tight control over all external advertising (including site boards promoting any development; see Chapter 8). This is frequently a matter in respect of which the planning authority will have a policy document. Normally the greatest weight is given to considerations of visual amenity and public safety in those cases where advertisements might possibly distract the attention of road users.

5.8   THE DEMOCRATIC ASPECT

Some types of application (normally minor in nature) might be decided by the planning officers of a planning authority using powers delegated to them by the planning committee. Lesser items, such as house extensions or advertisements, might be dealt with quickly in this way. More important or substantial applications will be decided by the Planning Committee of elected representatives of the authority, who will meet regularly and normally have full powers to make a decision on the application on behalf of the council. The committee will be advised by planning officers whose advice is based upon their knowledge of the area and its problems and policies, and upon consultations they have had with such bodies as the highway and water authorities. Their professional advice (which takes the form of a report) will also be influenced by their knowledge of planning case law and how the council's policies operate.

Normally the planning committee will heed the advice of their professional officers and will also take note of any representations made by members of the public. However, it is important to note the committee is not obliged to accept the officers' recommendation and may make a decision against the recommendation. Development is often highly contentious and public comment might come to the committee from individuals or groups or in the form of petitions. The committee must consider the variety of advice and representation it receives and make a decision to approve or refuse the application or perhaps, in some cases, to refer it back to the applicant to seek a modification.

Public interest in planning is often strong when the existing status quo might be disturbed. The planning committee is a committee of elected councillors who, as part of the democratic political machinery, recognize the impact that the decisions they may make has on their political standing. It is important for developers to appreciate, particularly where large controversial schemes are being prepared, the political make-up of a specific planning authority, in practice, can be a major influence on the eventual decision made on the planning policy issues.

5.9   THE PLANNING CONSENT

Planning permission may be qualified in various ways and it is proposed briefly to consider here the more important of these. Permission may be granted subject to a variety of conditions, for a limited duration of time or for the personal benefit of certain people or organizations.

With regard to time limits, there are two separate aspects to be considered. Every detailed planning permission will lapse unless the development is commenced within 5 years from the date permission is granted or such other time as the planning authority may specifically stipulate. In the case of outline permission, the necessary application for approval to the various reserved matters must be made within 3 years of the outline consent being granted. The development itself must commence within 5 years of the date of outline approval or within 2 years of the date of detailed planning approval, whichever is the later, subject to the imposition of any other specific time limit by the planning authority. Only a very limited amount of work on site (known as 'material operations') is necessary to prove that development has commenced to meet the time-limit requirements. The Courts have decided that in order to show that a planning permission has been implemented, there must be a genuine intention to carry out the development concerned when a commencement is made on the site. This 'test' has caused developers problems, particularly as planning permissions obtained at the height of the property boom in the late 1980s were not implemented and were due to lapse in the slump of the early 1990s. Many developers choose to renew permissions (which often requires renegotiations with the planning authorities) or in some cases let existing consents lapse rather than commence work on site.

To deal adequately with the evasion of the time-lapse provisions attached to planning permissions, local authorities currently have powers (via Section 94 of the 1990 Act) to serve a completion notice, the effect of which is that planning permission will lapse unless the development is completed within a reasonable period of time. Completion notices must be approved by the Secretary of State and there are rights of objection for the people who are affected by them which are considered by a Department of the Environment Inspector.

The second and quite different aspect of time limits is when the planning permission remains in force for a limited period, and at the end of that period any buildings or works which have been erected must be removed and any use authorized by the permission must cease. Thus at the end of the limited period of planning permission things will revert back to the state which existed before the permission was granted. On the expiry of limited period planning permission, it is of course open to the applicant to make a new application for the retention of the buildings or the continuance of the planning use. Planning permissions for limited periods are obviously of limited value.

Conditions may be imposed limiting occupation to a particular type of occupier. These fall into two broad classes. There is the condition which limits occupation of a building to someone engaged in some particular trade or vocation – perhaps the best-known condition of this type is limiting the occupation of agricultural cottages to those engaged in agriculture. The other and more restrictive type of condition, although rare, is one which limits occupation to a particular occupier personally. The ability to impose limitations on the type of occupiers will sometimes enable a planning authority to grant a permission which otherwise it would not be prepared to consider.

The powers of planning authorities to impose conditions on a planning permission are set out by government advice in the Department of the Environment Circular 1/85, which sets out 'Model' conditions relating to various aspects. The courts have laid down that conditions should be imposed only where they are necessary, relevant to planning, relevant to the development to be permitted, enforceable, precise and reasonable. A typical example might be a condition requiring landscaping proposed as part of a scheme, to be carried out in the next planting season available and maintained thereafter to the satisfaction of the local planning authority. There is the same right of appeal against the imposition of a condition on a planning permission as there is against the refusal of planning consent.

5.10   PLANNING AGREEMENTS OR OBLIGATIONS

Whilst the planning authorities prepare the necessary policies and plans to establish the planning framework for a particular area, this in itself does not bring about the implementation of the development plan. Implementation depends upon landowners and developers who initiate development proposals within the planning area. Having prepared the planning framework, the local planning authority normally waits for developers/landowners to appear and produce specific planning proposals and to make planning applications. In addition to the power to, where appropriate, attach conditions to the grant of permission it is possible for planning authorities to enter into legally binding agreements with developers which enable development proposals to come forward in circumstances where the authorities could not rely solely on their statutory powers of control, i.e. planning conditions. Planning agreements (referred to as planning obligations) between a developer and a planning authority may be made under the provisions of Section 106 of the Town and Country Planning Act 1990. Agreements might be made to phase the development of land to accord with the dates when various public services will become available or improved road access will be provided. Similarly, agreements might be made with regard to the provision of land within a comprehensive development area for public open space or amenity purposes. Where there is inadequate infrastructure, e.g. a lack of main sewers or adequate road access, it might be possible for a Section 106 agreement to be entered into requiring the developer to make financial contributions towards the cost of making available the infrastructure such that development may proceed. With regard to the provision of infrastructure, it should be remembered that agreements may be entered into with other authorities (relating to the adoption of roads and sewers, etc.) as well as the local planning authority.

If developers feel that a local planning authority is attempting to exert undue pressure on them to enter into an agreement which will impose unduly onerous burdens, their remedy is to make a formal planning application and take the matter to appeal if planning consent is not granted. However, in many cases it will be to the advantage of developers to offer a planning obligation themselves if they consider that an appropriate contribution of a facility or infrastructure will enable the development to take place at a very much earlier date than would otherwise be the case. Since the 1991 Planning and Compensation Act, developers may offer a 'unilateral undertaking' to be bound by such an obligation. This can be considered as material by the planning authority in determining a planning application and at an appeal to the Secretary of State following a refusal of consent due to failure to agree terms on an agreement with the planning authority.

It must be remembered that agreements bind landowners and their successors in title and are registered on the legal title of the land affected.

There has been an increasing use of planning agreements, particularly involving local authorities with infrastructure burdens. Contributions towards the provision of infrastructure, whether they be given in cash or by way of sites for various public authority purposes, will obviously be reflected in the amount which a developer will be prepared to pay for the land needed for the development. However, very often the developer may have to assess the likelihood of a Section 106 agreement and its cost before a site is purchased which adds considerable uncertainty.

The extent to which a planning authority should, when considering a planning application, negotiate with a developer in order to obtain some material benefit for the community (referred to as 'planning gain') has always been a matter of some controversy.

Department of the Environment Circular 16/91 provides guidance on the circumstances in which Section 106 agreements are considered to be reasonable. The guidance confirms that Section 106 agreements should only be sought where they are necessary to the granting of permission, relevant to planning and relevant to the development to be permitted. It states there are four tests of reasonableness. First, the agreement is needed to enable the development to go ahead, e.g. provision of road access (which can be a financial contribution), or it is so directly related to the proposed development that the development ought not to be permitted without it, e.g. the provision of public open space. Second, the extent of what is required is fairly and reasonably related in scale and kind to the proposed development. Third, what is required from the developer enables the implementation of local plan policies for a particular area or type of development, e.g. affordable housing. Fourth, what is required mitigates the loss of or impact on amenities or the existing environment, e.g. a contribution to enable nature conservation. The provision of social infrastructure by a developer is considered a legitimate requirement.

Some feel that it is wrong for a planning authority to bargain to obtain a material benefit in return for a planning permission and that such a practice brings the whole system of planning control into disrepute. Others feel that developers should not be subjected to *ad hoc* demands from planning authorities which are regarded as potentially dangerous precedents in the nature of local taxes on development. However, there are many who argue that, used in moderation and with common sense, planning gains can often facilitate development in circumstances where the authority is not able to provide needed facilities and the planning gain makes a contribution to the welfare of the community in which the development takes place. It must be remembered that the applicant has a right of appeal if planning consent is refused for any reason, and where a developer has submitted a unilateral obligation to enter into a Section 106 agreement on specific terms, the nature of the those terms and their appropriateness in the context of the Circular 16/91 tests is examined in detail.

5.11 BREACHES OF PLANNING CONTROL

Local planning authorities have wide powers to ensure as far as possible that no development which requires planning permission takes place; that no unauthorized uses are allowed to continue unless the planning position is regulated and that all development permitted is carried out in accordance with the conditions which the authority has imposed to the permission. These powers are included in the 1990 Act and summarized in their Policy Guidance Note.

In cases where development has allegedly taken place without permission the authorities are empowered to firstly require information concerning the development, the identity of the owner of the land involved and other pertinent matters to be divulged to them. They can serve an enforcement notice on the landowner/developer which identifies the breach of planning control, the action required to remedy the breach and a time limit to carry out the necessary action. There is a right of appeal against the notice on one of seven grounds. The most commonly used ground is that planning permission should be granted for the development concerned and the lodging of an appeal on that ground is regarded by the Department of the Environment (who consider the appeal) as being a deemed planning application. Other grounds include stating that the alleged breach has not taken place and that the means required by the planning authority to rectify the situation are unreasonable.

Additional powers are available in the form of breach of condition notices, where authorities can require that conditions on a planning permission are complied with. There is no right of appeal against such a notice, as the notice is then pursued to the Magistrates' Court via a criminal prosecution which can be defended. More immediate powers of action are contained in a stop notice which literally requires a specific activity to cease, normally as a forerunner to enforcement action being taken. If an appeal is successful against enforcement action where a stop notice has been served, there is a compensation procedure.

5.12   PLANNING APPEALS

In the event of the local authority refusing planning permission for the development proposed, or in the event (subject to the discretion of the applicant) of the authority either taking longer than 8 weeks to determine the planning application or granting permission subject to a condition which aggrieves the applicant, there is a right of appeal to the Secretary of State for the Environment. To be valid, an appeal against a refusal of planning permission must be lodged within 6 months of the date of refusal. In deciding whether to lodge an appeal against a refusal of permission, an appellant must be satisfied that there is sufficient evidence to suggest that the local planning authority's reasons for refusal are inappropriate. Planning authorities are required by central government to ensure that reasons for refusal are sound and clear-cut, and the onus at the appeal is placed on local authorities to demonstrate that the refused proposals would cause demonstrable harm to interests of acknowledged importance. In assessing whether reasons for refusal are capable of being set aside on appeal, the appellant (and their advisers) would examine the relevant planning policies (particularly in the context of Section 54A of the 1990 Act) and any other material considerations.

Once a planning appeal has been lodged, jurisdiction of the case passes to the Department of the Environment's Inspectorate. This is a body of qualified inspectors appointed by the Secretary of State to determine planning appeals on his behalf or, in certain cases, to make recommendations to the Secretary of State to enable determination of the appeal himself. There are three alternative methods of appeal which can be pursued by appellants.

5.12.1 Written representations

This is an exchange of written statements with the appellant's statements putting forward the case in support of the appeal being allowed and the local authority's statement in reply seeking to justify the reasons for refusal. Both statements are considered by the Department of the Environment Inspector, who visits the appeal site and determines the appeal on the basis of the written evidence and the site visit.

5.12.2 Informal inquiry

This takes the form of a meeting between the appellants and the local planning authority with the Inspector acting as chairman. Written statements of case must be submitted to the Department of the Environment in advance of the hearing. Having read the statements, the Inspector usually identifies those issues which he/she requires to be discussed at the commencement of the hearing and these issues are aired between the parties. Following the conclusion of the hearing, the site visit will be held and the Inspector will again determine the appeal on the basis of the written statements, the discussion at the hearing and the site visit.

5.12.3 Formal public inquiry

This is a quasi-judicial inquiry where the appellant is represented by a barrister, solicitor or qualified person who presents the case on behalf of the appellant, calling expert witnesses to give evidence in support of the appeal being allowed. The planning authority is similarly represented, usually by the council's own solicitor, who again calls expert evidence supporting the refusal most often from the council's Planning Officer. The witnesses are cross-examined by the opposing party and the Inspector, having listened to the evidence presented at the inquiry and following the site visit, then makes his decision or recommendations to the Secretary of State.

The three methods of appeal are the subject of detailed guidance on procedure through Department of the Environment Circular 10/88.

The Spindles, Oldham – a 280 000 sq ft (26 013 sq m) town centre retail scheme developed by Burwood House Developments Ltd which opened for trading in September 1993 and is currently 90% let. This photograph shows the two-level main shopping mall within the scheme.

Department of the Environment Circular 8/93 provides for costs to be awarded against either party at an inquiry if there has been unreasonable behaviour. This definition normally applies to local planning authorities who have unreasonably withheld planning permission for proposals that were clearly acceptable or where grounds for refusal have been withdrawn at a late stage. Until relatively recently costs were infrequently awarded against appellants. However, the government has made it clear that, given its commitments to Green Belt policy and the development plan, in certain cases, therefore, where such proposals have been brought forward by developers to appeal (which are clearly contrary to an adopted development plan and where there are no material considerations suggesting the plan should be followed), costs are to be awarded against the appellants.

With the increase in economic activity in the property boom of the late 1980s and the number of planning applications which were refused, the number of appeals rose so that in 1989–90 the number of appeals considered by the Department of the Environment in England and Wales was 34 487 (Department of the Environment Statistics, 1990). Of the appeals heard in 1989–90, 36.7% were allowed (Department of the Environment Statistics, 1990). With the impact of the recession the level of activity declined. In addition since the 1990 Act was passed and the emphasis was placed more on the development plan, developers have been advised to pursue schemes in the context of their involvement in the development plan process outlined earlier as a means of facilitating a favourable policy framework in which to submit an application. This twofold change, one economic and one legislative, has meant the number of planning appeals has declined in England and Wales to 16 038 in 1993–94 (Department of the Environment Statistics, 1994) but the success rate has remained the same. This is a reflection of the government's intention for the planning system to be 'plan led'. The time taken for the process to run its course, given the need for evidence to be presented and considered by the Inspectorate, has become somewhat lengthy. It is normal for the public inquiry appeal route to take between 6 and 9 months to be completed from the date of lodging an appeal in receipt of a decision. In particularly busy local authorities where a backlog of appeals by public inquiry may have built up, the process is known to take much longer. The time taken for appeals by written representations and informal inquiries are somewhat shorter, but the delays which occur in the decision-making process are clearly a factor to consider when making a planning application which is likely to go to appeal.

5.13   THE FUTURE

In that the planning system is slave to the political and legislative system, the future is difficult to predict. The development industry is generally dissatisfied with the delays and uncertainty which it experiences and sometimes the *ad hoc* nature of the decisions made, particularly where local politics holds sway over professional logic. Various proposals have been put forward to improve the system, including a proposal whereby developers pay additional planning fees to give their particular application priority treatment.

In a similar vein the Conservative government's eagerness to involve the private sector in what has historically been a public sector orientated decision-making process is starting to manifest itself in private companies carrying out certain tasks for planning authorities. These include the preparation of local plans, the carrying out of studies into the impact of retail proposals and advice on the development of key sites. That trend appears to be one which will continue in the current political climate.

The Labour Party's views on the planning system are as one would expect, less geared to the private sector. Amongst their ideas are to abolish the general right of appeal against unfavourable planning decisions and to make the development plan more dominant than the current 1990 Act suggests.

The future direction of the planning system and how it affects the development industry depends on factors outside its control, although it has certain key groups who lobby on its behalf – such as the British Property Federation, the Royal Institution of Chartered Surveyors and the Building Employers Confederation. The long-term uncertainty on how the system itself will change is therefore added to short-term uncertainty relating to how each individual application will be dealt with.

Executive Summary

The planning system in England and Wales is based upon decision making at three levels of government with particular emphasis placed on the district councils. For most forms of development involving householder applications and minor changes of use, the system generally works in a straightforward manner. However, for more major forms of development, the political nature of the system commonly means that the simple exercise of completing an application form correctly is insufficient to obtain planning permission. Consultation between developers and council officers, local interest groups, highway authorities and others on anything but the smallest project is now regarded as critical – and this, combined with the increasing role of the development plan and its adoption process, has changed the way in which developers regard this system.

In a general book on property development the broad principles set out above provide a good basis upon which to approach the planning system. It should be noted, however, that like valuation, tax or building surveying, planning matters have become more of a specialism in which developers often employ specialist advice. The heightened concern of government and others in relation to environmental matters has greatly influenced the planning system. This concern has placed town planning in its widest sense in a high place on the political agenda.

REFERENCES

Ball, S. and Bell, S. (1991) *Environmental Law*, Blackstone Press, London.
Moore, V. (1993) *A Practical Approach to Planning Law*, 3rd edn, Blackstone Press, London.

6

CONSTRUCTION

6.1 INTRODUCTION

After the acquisition of the land, the developer's second major financial commitment is the placing of the building contract which finally commits the developer to a particular building design and content. It might be regarded as the point of no return with much of the developer's earlier flexibility gone. The developer's aim in this stage of the development process is to produce a good quality building on time and within budget.

In this chapter we shall firstly consider the various types of building contract. Secondly we shall examine the management of the contract from the necessary pre-contract procedures through to the handover of the building and defects liability period. We will comment on the role of the professional team and the project manager during this phase of the development process. Also we shall examine the various risks inherent in any building contract and how they may be reduced or shifted from the developer to the building contractor or vice versa.

The successful outcome of any development will be greatly influenced by the efficiency with which the building arrangements are made and carried out.

The type of building contract to be adopted for any particular development must be decided at an early stage. This decision will significantly influence the size and composition of the development team. As soon as the brief and schedule of accommodation has been prepared and the broad design constraints decided, the decision on the type of building contract must be made. There are various possible arrangements, and the extent to which developers, albeit through their professional advisers, are responsible for the building design and management of the building contract, and the degree of risk to which they will be subject, varies according to the particular contract arrangement.

6.2   CHOICE OF BUILDING CONTRACT

The developer's decision on the form of contract will depend on their requirements and the size and complexity of the development. Developers must decide whether the design of the building is to be carried out by an architect and the professional team, or the contractor. There may be a need to shorten the pre-contract time by overlapping the design and construction elements of the scheme. Alternatively, an early completion may be achieved by using fast-track methods of construction. Of crucial importance in deciding on the form of contract is the likelihood of changes to the design during the contract and the need for flexibility. An additional factor is the extent to which the developer wishes to pass risk onto the contractor. A public sector developer will be also concerned to achieve value for money to meet the requirement for public accountability.

There are no hard and fast rules about choosing any particular form of building contract. A developer may use any of the types of contract available and adapt them to suit their own particular requirements, provided that it is generally acceptable to building contractors. However, there are advantages in using one of the forms of contract arrangement familiar to those in the building industry, and in respect of which there is a good deal of practical experience, so that the strengths and weaknesses of that particular type of contract are known. From the viewpoint of the developer, building contract arrangements (often referred to as procurement methods, i.e. methods used to both design and construct a scheme) may be broadly divided into three main categories, albeit with many variations in each.

The first is based on the use of the traditional standard form of contract evolved by the Joint Contracts Tribunal (JCT), which provides for a main contractor to carry out the construction in accordance with the designs and specifications prepared by the developer's own professional team – and upon whom they must rely for the quality of design, adequate supervision of construction and suitability of the building for the purpose for which it is designed.

The second is the 'design and build' contract which is completely different and is increasingly being used in preference to the traditional contract. The contractor is responsible not only for the construction, but also for the design and specification, and must take full responsibility for ensuring that the building meets the requirements of the developer and is fit for the purpose for which it is designed.

The third is management contracting, based on American methods of construction, which was used by some of the larger development companies on complex developments in the 1980s. Like the JCT contract, the professional team is responsible for the design and specification. However, the building work is split into specialized trade contract packages and the management contractor – for a management fee – co-ordinates and supervises the various subcontractors on behalf of the developer.

6.3   THE JCT STANDARD FORM OF CONTRACT

This contract is drafted by the JCT, which comprises a number of bodies – the Royal Institution of Chartered Surveyors, the Royal Institute of British Architects and the British Property Federation (representing developers and property owners).

This contract has been subjected to a good deal of criticism, particularly in recent years; nevertheless it continues in use. It tends to be used on straightforward small to medium sized schemes. A number of variations and amendments are frequently used but the following describes a typical contract.

Developers appoint their own professional team who are responsible for the design of the building to meet their requirements, for supervising the construction process and generally administering the contract. The architect together with the project manager (if appointed) lead the professional team and call in whatever other team members they need to deal with such matters as structural design problems and the provision of mechanical and electrical services. Quantity surveyors should be appointed at the outset, not only to provide a cost estimate but to provide cost-planning services. We shall now examine the role of the professional team in designing and administering the scheme under this type of contract.

6.3.1   The role of the professional team

The architect is responsible for obtaining planning permission (unless a planning consultant is appointed) and all other statutory approvals such as building regulation approvals and fire certificates. The architect is responsible for the design of the buildings in terms of aesthetics and functions in accordance with the developer's brief and budget. The architect also is principally responsible for the management of the contract, although supervised by the project manager if one is appointed. The architect does not supervise the building contract on site on a full-time basis. Accordingly, a developer may also appoint a clerk of works or resident engineer at extra cost to carry out a full-time 'on-site' supervisory role.

The quantity surveyor is responsible for preparing estimates of building cost, preparing Bills of Quantities (measured specification of materials and work to enable the contractor to submit a price) and, during construction, for preparing valuations of work upon which the architect issues the 'interim' and 'final' certificates. The quantity surveyor should be appointed as early as possible within the development process to advise on cost and the merits of alternative forms of construction (e.g. steel frame or concrete frame) and should also provide estimated cash-flows of the building contract expenditure. The quantity surveyor reports on the cost of construction and measures actual payments against the estimated cash-flow. Their role is to explain why the actual cash-flow differs from the estimate and prepare revised estimates for the remainder of the project. The quantity surveyor is also responsible for estimating the cost of possible variations in design so that the development team can decide whether or not they are merited.

Other members of the professional team may include structural engineers and mechanical and electrical engineers. The structural engineer works closely with the architect and quantity surveyor, in assisting with the design of the structural elements of the building, calculating loads and stresses, and advising upon how the design of the building should be modified to accommodate them. A mechanical and electrical engineer advises on all the services required such as electricity, gas, water and the design of the heating, lighting and plant installations, and where air conditioning is to be provided, they will also be responsible for designing the system and liaising with the architect, so that it can be incorporated into the design.

Provided that the contractor executes the building work in a good and workman-like manner and in accordance with the architect's drawings and specification in the Bills of Quantities, and with any instructions subsequently given to the contractor by the architect, the contractor will not normally have responsibility if the building is not suitable for the purpose for which it was designed. This is irrespective of whether the unsuitability is attributable to faulty design or to some physical inadequacy in the structure. Developers must turn to their architect and other professional advisers for a remedy. On occasions, the respective responsibilities of the professional advisers for an inadequacy or defect in the building are not clear-cut. In such circumstances developers find themselves in a somewhat impossible situation, dependent upon the outcome of the arguments between professional advisers. Such situations may be avoided by the exercise of care in the selection of the team of professional advisers, and where a highly complicated or specialized or sophisticated building is involved, it is helpful if the professional advisers have had previous experience of dealing with that particular building type. The quality of the team has a significant influence upon the success of the development project.

In most cases, all the professional team including any project manager, are appointed on a percentage fee basis, either by negotiation or in accordance with their professional body's suggested scale fee. The percentage fee will usually directly relate to the final building contract sum. From the developer's point of view this provides no financial incentive for the professional team to ensure that the building is built on time and within budget, although the professional's reputation is at stake along with their chances of being appointed again by the same developer. A developer may be able to negotiate a fixed fee basis for appointment. However, the developer must be aware that in such circumstances the professional concerned will invariably include an element in their fee proposal to cover the risk for extra work, and therefore the developer may not necessarily gain any advantage.

Developers will require the design team, i.e. architect, structural engineer, mechanical and electrical engineer, and sometimes the quantity surveyor, to enter into deeds of collateral warranty for the benefit of investors/purchasers, financiers and tenants. In effect, these warranties extend the benefit of the developer's contract with the professionals involved. Typically, they require the professional practice or company to warrant that all reasonable skill, care and attention has been exercised in their professional responsibilities and that they owe a duty of care. In addition, the professionals are required to warrant that they have not specified deleterious materials. They will also need to provide evidence of sufficient PI cover (professional indemnity insurance) from their insurance company. It is an often difficult and time-consuming process for developers to procure these warranties in a form acceptable to both parties, as financial institutions and banks require almost total responsibility from the professional design team. The professionals, in turn, are increasingly resisting deeds of collateral warranty in the form required by financial institutions and banks due to the restrictions placed on them by their insurers in relation to their professional indemnity policy. It can become very difficult for the developer acting as agent in the middle. For instance, professionals may refuse to sign these agreements under seal, that is they wish their responsibilities to last 6 years instead of 12. Additionally they may wish to limit the assignability of their deeds of collateral warranty to the first purchaser and first tenant of the completed development. They also insist their liability is restricted to remedial costs of any defects, not consequential or economic loss. It is important, at the very least, for the developer to ensure that every member of the design team signs a warranty before they are appointed.

Recently, some developers have turned to insurance in the form of latent defects insurance known as decennial insurance, particularly on the larger schemes. This form of insurance is expensive, typically 1–2% of the building contract sum, but it has the advantage that the insurer is responsible for pursuing remedies with the professional team. The insurer assumes responsibility for repairing the property should an inherent structural defect be discovered which renders the building unstable or threatens imminent collapse. The insurer usually agrees to cover a project for up to 10 years from completion, provided that an audit has been carried out before construction begins and an independent engineer reports on the design and construction of the building. The developer will have to pay for the fees of the independent engineer. The insurance policy is totally assignable to tenants and purchasers, who seem to be increasingly demanding such insurance policy is in place (see the Case Study in Chapter 4 as an example)

The majority of the design work on this type of contract is carried out prior to the appointment of a contractor and has a considerable influence on the final cost. Once the contractor has started work, changes necessitating revised instructions to the contractor usually result in an increase in costs or delays to the contract programme. Revised instructions are mainly caused by either the developer making a change, or the architect issuing late instructions as the design is inadequate or incomplete. Therefore, the relationship between the developer and the architect is crucial at the design stage. At the outset, developers should establish positive and realistic cost limits, which should be relayed not only to the architect but the whole professional team. They should ensure that the architect thoroughly understands – preferably with the aid of a written brief – their requirements in respect of all the aspects of the building and its usage, the standard and type of finishes required, the services needed and date for completion. Above all they should avoid if possible changing their mind unless necessary to secure a particular tenant or improve the value of the scheme. It is essential that the architect is chosen with great care as much relies on the ability of the architect to produce the working drawings for the contractor on time. Therefore, the developer should question whether a particular architect has the capability and resources to handle the design element efficiently.

6.3.2   Choosing the contractor

Once the detailed design by the professional team is complete then the quantity surveyor prepares the Bill of Quantities, which specifies and quantifies the materials and the work to be carried out in great detail – right down to the number and make of door locks. Then the procedure is to invite contractors to submit tenders for carrying out the contract work based on the detailed drawings and Bill of Quantities. However, this is not essential: there is nothing to prevent one contractor being asked to price the Bill of Quantities. This might occur when the developer has been used to employing a particular contractor (it may be an in-house contractor) over a period of time and is entirely satisfied with the result of the contractor's work and, for that reason, prefers to employ the particular contractor again. It may be that the contract contains a great deal of highly specialized work for which one contractor has an outstanding reputation and might be chosen on that ground to carry out the work subject to a satisfactory pricing of the Bill of Quantities. If this particular route is taken then it may be advantageous to appoint the contractor earlier in the development process to advise the design team on the practical aspects of the design.

When choosing contractors to be invited to submit competitive tenders for carrying out the work, there are various considerations to be borne in mind. In practice, it is necessary to limit the total number of contractors invited to submit competitive tenders. Six or eight contractors are often adequate for even the largest contract, because the pricing of Bills of Quantities for a large job can take a considerable amount of time (and therefore money) on the part of the contractors who are not keen on submitting competitive tenders when there is, in their opinion, an unreasonably large number of tenderers.

If the work is of a specialized nature, contractors skilled in that particular type of work are obviously chosen. On occasion, it is thought advisable to employ a large national contractor, while sometimes local or regional contractors are preferred. Some contractors have a particular reputation for producing work of high quality, others for producing work quickly and on time, and this can be of considerable importance to the developer's cash-flow. Still other contractors have a reputation for submitting keen tender prices, and there are those who have a reputation for their considerable expertise in formulating claims for extra payments on any and every occasion during the contract. The architect or developer may feel that some contractors are entirely dependable, while others may have let them down on a job in the past. Unfortunately, high-quality work and speed and low cost are a very rare combination. Developers are normally guided by the advice of their architect, project manager and quantity surveyor on the selection of the contractors for the tender list. Before any contractor is included, they should be asked whether they are willing to tender for the particular job. There are times when contractors are fully extended and are disinclined to tender for extra work. Some contractors for a variety of reasons may not be interested in tendering for work in a particular locality.

The prices submitted by each of the contractors will be examined carefully by the quantity surveyor and the project manager to establish what is being offered. The quantity surveyor will independently price the Bill of Quantities and this will then be used for comparison purposes with each of the contractor's tenders. Each contractor will price each element of the work, setting out the applicable rates for each element of work or unit of material. The contractor's priced Bill of Quantities forms part of the contract documentation and will be used by the quantity surveyor to value the work carried out by the contractor.

The reliability and financial stability of the contractor are important considerations. Therefore, when contractors have been chosen, it might be thought desirable for them to take out what is known as a 'performance bond'. The contractor usually takes out a performance bond with an insurance company which guarantees to reimburse the employer for any loss incurred up to an agreed amount as a result of the contractor failing to complete the contract. Financiers may also insist on a performance bond for their benefit. The failure of a contractor is a major disaster from the developer's point of view. Very often there are long delays while the legal position is sorted out and another contractor found who is prepared to complete the work. The new contractor might ask a considerably higher price for completing the job than was contained in the original contract. If defects subsequently appear in the completed building, it might be very difficult to apportion responsibility as between the original contractor and the contractor who takes over and completes the job. Thus it is easy to understand why many employers ask for a performance bond, even though the extent to which their losses are reimbursed is limited and the cost of the performance bond is normally added to the contract cost.

6.3.3   Paying the contractor

The method of payment for the building works has a considerable impact on the developer's cash-flow position. It also has considerable impact on the contractor's cash-flow who has to have regard to the method of payment when preparing the price for the work. For that reason, the method of payment must be made clear at the time the contractor is invited to tender.

Under the JCT contract, the architect typically authorizes monthly payments based on the value of work certified by the quantity surveyor. Usually a certain percentage (3 – 5%) of the total value of the work carried out is retained until the end of the contract, which is known as the retention. This type of arrangement best suits the contractor who obtains payments for the work carried out irrespective of when the building is ready for occupation. The developer has to pay out very substantial sums of money over a considerable period of time before obtaining the benefit of a completed building at the end of the contract.

The ideal arrangement for developers is for the whole of the contract price to be paid when the building is handed over, so that they do not part with their money until the time when they should be receiving an income from the building or have the benefit of occupation of it. It must be remembered, however, that if it were possible to make such an arrangement (it is extremely rare), contractors would increase their tender prices by one means or another to take account of interest and the additional risk. In the case of a large contract spread over a period of time, some contractors might not be able to finance the work easily without payments from the developer. Some compromise might well be devised for payment to be made in certain set stages – the last payment on completion and handover of the building being heavily weighted so as to give the contractor an incentive to get the building finished. The method of paying for the work obviously has to be related to the particular circumstances of each contract. The contractor is more likely to be flexible if they are a partner in the scheme and stand to benefit in terms of a profit share.

6.3.4   Calculating the cost

When using the JCT form of contract, it is usual for the contractor to submit their bids on either a 'firm price' or a 'fluctuations' basis. The 'firm price' means that although the cost of labour and materials used in carrying out the work may fluctuate with the market, the contract sum will not be varied to take account of these fluctuations, whereas the 'fluctuations' basis means that once the contract has been awarded to the contractor any increase or decrease in labour and materials will automatically be added to or subtracted from the contract sum. Under both types of contract, the standard form contains a clause allowing adjustments to be made to take account of alterations in cost due to government legislation. However, the developer may delete such clauses, particularly when the tender market is very competitive (as it is at the time of writing). It is important to stress that 'firm price' does not mean the contract sum once fixed will not alter. Quantity surveyor's remeasurements, architect's variation orders and instructions and extensions of time may affect the cost.

Developers and their professional advisers must decide on what basis they wish contractors to prepare their competitive bids in order that the contractors submit prices on the same basis. Developers do not always find it easy to decide which basis is likely to be to their advantage. The risk of fluctuations in the cost of labour and materials during the contract is a fact of life and cannot be avoided; the question to be decided is whether the risk is to be borne entirely by the contractor as in a firm price contract, or whether the risk is to be borne by the developer as in a fluctuations contract. If contractors have to prepare their bids on a firm price basis, they will clearly add something to their prices to cover themselves against the risk of increased costs. Very often to tackle this problem contractors are asked to quote prices on both a firm and fluctuations basis; leaving the developer to decide in the light of the differing prices which basis is likely to prove most advantageous during the whole of the contract period. Each contract must be judged on its own merits against the background of tender market conditions.

There are two main alternative methods of calculating the cost. Firstly, what is known as a cost-plus contract may be entered into. The contractor is paid on the basis of the actual cost of the building work ('prime cost') plus a fee to cover their overheads and profit. The fee might be fixed or a percentage fee calculated with reference to the final building contract sum or the initial estimate. Secondly, a target cost contract might be negotiated or established by tender. A target cost is agreed with the contractor plus the contractor's fee. Any savings or additions to the target cost are shared by the parties.

6.3.5   The duration of the contract

The date agreed in the contract for the completion of the building is not certain as the contractor can apply for extensions of time for a number of reasons. Some of these reasons which justify an extension of time also entitle contractors to recover any additional loss or expense they may have suffered as a result of that extension. Extensions of time almost always lead to increase in the cost which will certainly include the contractor's 'preliminaries' (overheads such as insurance, cost of plant hire, etc.) The impact of an extension is felt twice by the developer: firstly, it effects the cash-flow and, secondly, it increases the costs.

The main reasons for extension of time which entitle the contractor to recover additional loss and expense are:

1. Inadequacy in the contract documents: the drawings and/or the Bill of Quantities. This may be attributable to the incompetence of the professionals responsible for the preparation of the documents or new legislation might be introduced during the contract which requires some amendment to the drawings. Additionally, the Building Control Officer of the local authority (responsible for providing building regulation approval) or the local fire officer (responsible for issuing a fire certificate) might impose conditions on their approval which necessitates some design changes. Unforeseen ground conditions (in the case of a cleared site) or hidden structural problems (in the case of a refurbishment) can also cause additional expense and delay. This underlies the necessity for a thorough site investigation and/or structural survey before the tender documents are prepared.
2. Delay by the architect in issuing drawings or instructions.
3. Delays caused by tradesmen directly employed by the developer.

Additional reasons which may be included in the contract terms (depending on the results of negotiation with the contractor on the standard clauses), entitling the contractor to an extension of time, but not to recover any additional loss or expense, include:

4. Failure by the nominated subcontractors: on almost every building contract some of the work is subcontracted to specialists. When contractors find and appoint their own subcontractors, they are entirely responsible for any delays caused by them, so that the developer does not suffer as a result. Architects often nominate particular subcontractors to carry particular elements of the building. Among the reasons for doing this may be the high quality of the subcontractor's previous work with the architect, or a particular expertise in both designing and constructing elements of the work, e.g. the structural steel work.

5. Bad weather: architects should ensure a careful record is taken of weather conditions on site.
6. Strikes and lockouts.
7. Shortage of labour or materials.
8. Damage by fire where it is the contractor's responsibility for insurance under the contract.
9. Force majeure, i.e. acts of God such as earthquakes, floods, storms, etc.

Developers may be compensated for a delay for which the architect has not granted an extension of time under the terms of the contract. The compensation is in the form of liquidated damages at a rate previously agreed in the contract to cover the developer's loss. However, the agreed rate usually falls short of the developer's true loss in terms of the overall development cash-flow.

A developer cannot assume that the building work will be carried out within the time set out in the building contract and for the exact contract sum. One of the main advantages of the JCT contract is its flexibility with regard to the way the price for the job is to be fixed and its elasticity, which enables the type and quantity of work within the contract to be varied and yet leave the quantity surveyor free to negotiate the final price for the job at the end of the day. However, there is also a disadvantage with regard to the flexibility of this contract: it does not discipline the developer into making early clear-cut decisions as there is always the scope to make late variations. Also the competence and efficiency of the professional team is so important to the successful outcome of the contract. The inadequacies of the plans and specifications do not usually reveal themselves until the building contract is underway. The professional team is not motivated enough to control costs and delays: quite the opposite, their fees usually increase in proportion to the final contract sum. This type of contract may also lead to a very confrontational situation with the contractor if the tender documents are inadequate or many variations are made. Also, many contractors have a tendency to use the claims procedure as a negotiating ploy to claim additional money to cover losses they have made on the contract. It is for this reason alone that many developers are now using design and build contracts.

Developers must ensure that they are in control and are kept closely advised as to the likely financial outcome of the contract at all times. The best arrangements for dealing with this are examined in Section 6.6 on project management.

6.4   THE DESIGN AND BUILD CONTRACT

This is of a radically different nature: one party, usually the contractor is responsible for the design and construction of the scheme. In recent years its use has become more widespread due to dissatisfaction with the traditional JCT contract, and the problems encountered with splitting design and construction responsibilities. It has been traditionally used on simple and straightforward schemes but is increasingly being used on most types of building. It is particularly widely used by public sector clients for hospitals and schools, etc. The standard contract used for design and build is the 'JCT 1980 With Contractors Design'. The case study at the end of this chapter examines the variation of a standard design and build contract on a retail park.

The contract is based upon the production of a performance specification by or on behalf of the developer. 'Performance' means in this context the various requirements which the building must meet. However, the developer's requirements must clearly, and in as much detail as possible, be set out in the performance specification if this type of contract is going to work from the developer's viewpoint. The responsibility for the preparation of the design or what precautions should be taken to ensure that the finished building meets with all the various statutory requirements and suitability for the purpose for which it is designed must rest entirely with the contractor.

The performance specifications will vary from being fairly simple to very detailed depending on the nature of the scheme. For example, a developer might wish to develop a site by putting up some very simple standard-design warehouse units. The specification in this case may include a schedule of floorspaces for units of different size, with an indication as to how much office accommodation is to be provided, the total amount of toilet accommodation, the services to be put into the building, the floor loadings and the clear floor heights, together with an indication of the total yard area. On that simple performance specification, contractors would be asked to submit schemes for the erection of their own standard-design units to meet the requirements, together with their price for carrying out the contract. The complete responsibility for obtaining all the necessary statutory approvals, and for designing and erecting the buildings and ensuring that they will be suitable for the purpose for which they are required, rests with the contractor.

However, the performance specification tends to be extremely detailed and will specify the materials to be used by the contractor. It is also the responsibility of the contractor to comment on any materials specified by the developer if they consider them to be totally unsuitable for the purpose, before the contract documentation is entered into.

It is quite possible to arrange for complicated buildings to be erected under a design and build contract. However, in such cases, the performance specification is critical and has to be carefully prepared, probably by a team of professional advisers. In the case of a complicated building, which does not conform to a standard design, developers are entirely dependent upon the adequacy of their own performance specification to ensure that they get a building which meets their requirements.

A developer using this type of contract may appoint a specific contractor with which they have successfully worked before or who has a particular expertise in constructing buildings similar to the one proposed. Alternatively, a developer will go out to tender to appoint the most suitable contractor for the job based on their design, specification and price. Typically, the tender list will be short (two or three) due to the considerable amount of work involved by the contractor in preparing their submission. Obviously, as with the traditional JCT contract, the track record and financial stability of the contractor is an important consideration. Some contractors specialize in design and build, building up valuable experience in constructing and designing, while some of the larger contractors might have 'design and build' divisions. The contractor may employ all the necessary skills in-house or more usually employ external architects and engineers under the supervision of their own in-house project managers.

A developer may appoint a quantity surveyor to carry out the usual pre-contract activities such as advising on the most suitable form of contract and preparing initial cost estimates. The quantity surveyor will also perform similar duties during the contract as those on a traditional JCT contract, including valuations for interim payments and the final account.

In some cases, the quantity surveyor may take on the role as the developer's representative (referred to as the 'employer's agent' in the contract), and effectively project manage and administer the contract (an example of this is contained in the case study at the end of the chapter). This may involve the quantity surveyor in agreeing the letters of appointment and deeds of collateral warranty with the professional team, together with the chairing and minuting of all project meetings (usually the role of the architect under a traditional JCT contract).

Under a design and build contract the contractor submits drawings and specifications to the developer for approval, who can carefully check to see just what type of building will be built, what services will be provided, and so on. The contractor is responsible for designing and constructing the scheme in accordance with the approved drawings and specification. In a simple package deal where a standard type of building has already been erected by the contractor in a number of locations, and examples of which can be inspected and the occupiers asked for their comments on the adequacy of the building, developers will know the product and any feedback on its suitability. In such circumstances the design and build contract is particularly advantageous where the buildings are of a simple, straightforward nature, with often repetitive design elements, and can be built to a standard design which has been used by the contractor elsewhere. The advantages are that the design time and cost can be greatly reduced; the contractor is working to their own designs with which they are thoroughly familiar; and they may well use various standard types of component, which they can buy advantageously and which they are used to using on site. The contractor's own designs will undoubtedly reflect the contractor's practical experience of putting up buildings. The result should be that the contractor works more efficiently and speedily, and therefore more economically, so that the price of the building to the developer ought to be somewhat lower.

The advantages of this type of contract from the viewpoint of the developer are that while it is possible to provide for fluctuations in the contract price, and there are various alternative ways of paying for the buildings as the contract proceeds, usually a lump-sum fixed price is agreed: the contractor is committed to providing the buildings for a known cost and takes the risk involved in so doing. Obviously, contractors allow for these risks when preparing their price but the developer is greatly reassured to know that the price is fixed. It may only be changed by variations issued by the developer or changes in legislation. When the tender market is competitive (as it is at the time of writing) clauses in the contract dealing with changes in legislation may be deleted. The developer does not run the risk of becoming involved in endless professional arguments if the design or construction of the building is defective; it will be entirely the responsibility of the contractor to see that matters are remedied.

There are disadvantages, however. The developer does not have the same detailed control over the design (if that is required), and it is dangerous for the developer to request any alterations during the construction of the buildings because the cost might be increased out of all proportion – the developer does not have the protection of the flexibility of the traditional JCT-type contract. It might be argued that the final cost of the buildings under a fixed-price, lump-sum package deal might be higher in view of the risk which is carried by the contractor but, in practice, this is often offset by the advantages to the contractor of using his own design and standard components. Another important advantage is the likely achievement of an overall saving in time due to the overlapping of the design and construction processes. Also the developer will save money on professional fees as involvement by professionals will be considerably less.

There may be types of development for which the design and build type of contract is not suitable. Some developers believe that under this type of contract the aesthetics and quality of the finished building is lower compared to the traditional JCT contract. To overcome this problem a developer may appoint an architect to prepare the initial drawings and sketch designs under what is known as a develop and construct contract. The contractor is then responsible for developing the design as part of their tender submission. However, under this arrangement the developer must ensure that the design responsibility is adequately defined.

Alternatively (and this is increasingly the case), the developer may appoint an architect and engineer and novate their appointment contracts to the contractor (the case study at the end of the chapter provides an example of this arrangement). By novating the contract the contractor simply steps into the shoes of the developer and becomes the client of the professional concerned under the terms contained in the original contract. Thereafter, the contractor takes control of the design process and the professional is liable to the contractor. A potential conflict of interest may arise as the architect/engineer may continue to treat the developer as their 'employer' on the basis of their long standing relationship. It is important to ensure that the professional and the contractor develop a good working relationship. The developer will still insist on deeds of collateral warranty with each professional in case the contractor goes into receivership. The advantage to the developer under this variation is that they can appoint their own choice of architect and engineer, with whom they have established a good working relationship, whilst retaining the advantages of the standard design and build contract.

One of the main drawbacks of the design and build contract from the point of view of the developer (whichever variation is adopted) is the lack of flexibility. Developers must decide before the contract is signed on their exact requirements, as major variations can usually only be made at a considerable cost.

900 Slough, 950–959 Buckingham Avenue, Slough. Ten industrial/warehouse units varying from 5865 to 131 010 sq ft (543.85 to 1208.67 sq m) developed by Slough Estates plc and completed in 1994. The development has received a 'Very Good' rating under BREEAM (the Building Research Establishment Environmental Assessment Method) for its energy efficient design.

6.5   MANAGEMENT CONTRACTING

Management contracting, a method of construction developed in the USA, became more widespread in the UK during the 1980s. Developers were impressed with the fast track methods of construction used in the States. The management contract is generally used on fairly large complex development projects where the developer requires fast building times at competitive prices with the flexibility to change the design during the contract. Essentially the building contract is split into specialized contract packages, either by trade or building element, and let separately under the supervision of the management contractor.

The developer appoints the professional team in the usual way (as described in Section 6.3.2) to prepare the drawings and specification for the development project. The quantity surveyor prepares a contract cost plan based on the drawings and specification. The actual cost incurred by the management contractor ('prime cost') is paid by the developer having been certified by the architect and monitored by the quantity surveyor. The developer also pays a fee to the management contractors for their services in managing the various separate contracts. The developer may not necessarily appoint the contractor with the lowest fee proposal. It is important that the contractor has management contracting experience and sufficient staff with the right skills. The fee may be a lump sum or a percentage of the contract cost plan. The 'prime cost' includes the amounts due to the various contractors of the various parts of the project, plus the management contractor's own on-site costs. The construction work is carried out by the various contractors, who enter into a standard JCT contract, with the management contractor based on detailed drawings, specifications and Bills of Quantities. The selection of the contractors should be carried out by the developer and the professional team in consultation with the management contractor. The architect has the power to issue variations known as 'project changes'.

The management contractor should be involved at an early stage with the professional team in advising on the practical implications of proposed drawings and specifications, and the breakdown of the project into the various separate packages. The management contractor is paid in relation to interim certificates issued by the architect, including instalments of the management fee. The disadvantages are that the developer is having to pay the management contractor's fee, in addition to those of the professional team, and the management contractor is not responsible for the actual building works. The developer, therefore, has no direct contractual relationship with the various contractors carrying out the work. Accordingly, the developer must enter into design warranties with each contractor which are capable of being passed to purchasers, financiers and tenants. However, the management contractor may have to pursue the remedies of the developer in respect of any breaches by the various contractors. The developer has to reimburse the management contractor in settling or defending any claims from the contractors, unless the management contractor is in breach of the contract or of their duty of care. The liabilities and hence risk of the management contractor is limited. They are not responsible for the payment of any liquidated damages for any cost over-runs if caused by reasons outside their control or by delays due to the various contractors. It is essential that the management contractor, developer and the professional team co-operate since the developer is having to pay extra for the management contractor's expertise and experience.

It is important to remember that the management contract itself is not a 'lump-sum' contract. The management contract is based on the contract cost plan prepared by the quantity surveyor, which is only an indication of the price. However, the contracts the management contractor enters into with the various contractors are usually based on the standard 'lump-sum' JCT contract. Therefore, the final cost is based on the contracts with each specialized trade and it may bear no relation to the cost plan within the developer's contract with the management contractor. Cost control is essential under this type of contract and it must be very tightly managed by the project manager and quantity surveyor. There is no direct incentive for the management contractor to keep within the cost plan as their fee is directly related to the final building cost. The success of this type of contract in terms of cost control depends on the ability of the management contractor to appoint the various trades within the budget of the cost plan. However, the final cost is whatever the cost is to the management contractor (including the fee) and there is no penalty for exceeding the cost plan.

The developer has to accept a very high degree of risk with this type of contract. Developers who have used this type of contract have found cost control the biggest problem as there is no tender sum. There is no control over the delays caused by the individual contractors. There are extra costs involved in duplicating site facilities for the management contractor and the various contractors. However, the main advantage of management contracting is speed as projects are usually completed more quickly than on traditional contracts where full detailed drawings have to be prepared before the contract commences. This speed is achieved by the flexibility of dividing up the contract into separate elements, overlapping the design and construction of each element. The 'packaging' of the contract allows the developer to some degree to control costs and delays as contracts let later on in the process can be varied to suit. Pre-construction and construction times can be reduced when compared to other contracts. Developers are concerned that this type of contract does not effectively control the development's design and quality standards suffer as a result.

Due to some developers' bad experiences (due to the disadvantages mentioned above) with management contracting a variation known as 'construction management' was preferred towards the end of the 1980s. With construction management, the various trade contractors are placed directly by the developer and a construction manager is appointed for a fee as part of the professional team appointed at the same time, not necessarily afterwards like the management contractor. The most famous office development of the 1980s, Canary Wharf in London Docklands, used construction management. However, there are currently very few contractors available who have the capability or capacity to act as construction managers.

The construction manager really acts as the developer's agent and the appointment of a project manager is still required to provide a co-ordinating role with the professional team. Their fee is usually percentage based with an additional lump sum for the provision of site facilities. Their role is to manage and co-ordinate all the various contractors, review design proposals, control costs (against an agreed budget), control the contract programme and be responsible for quality control. The administration of claims for payment by the contractors and variations are their responsibility, although the final account with all the contractors is administered in the usual way by the architect and quantity surveyor.

The obvious advantage of employing a contractor on the professional team is to bring their experience and expertise into the design stage at the beginning. A contractor may employ 'value engineering' techniques (detailed studies of the cost-effectiveness of alternative materials and methods of construction) to review the design process. However, the developer needs to ensure that the contractor has the relevant design experience otherwise the advantage of strict design control will be negligible. The developer has to have much greater involvement in this type of contract arrangement and it is the job of the construction manager to ensure that developer makes firm decisions at the appropriate time. It is a very management intensive contract resulting in higher staff and fee expenditure on behalf of the developer compared to other types of contract.

The main advantage is the saving of time, achieved by overlapping the design and construction of each package and involving the construction manager at the beginning of the design process. It is appropriate where an early completion of the scheme is critical to the developer. However, despite the fact the developer has direct control over the various contractors, the problem of cost control remains. There is still no guarantee of what the final cost of the scheme will be, although incentives may be introduced to increase the contractors' share of the risk but at a cost to the developer. This method may be used by developers if they wish to maintain flexibility, take advantage of fast-track methods of construction and retain control while accepting greater risk. They could reduce this risk if they were confident of their exact requirements at an early stage and the pre-contract period was long enough to allow for detailed design.

6.6   PROJECT MANAGEMENT

The appointment of a project manager is not necessary for every project. A project manager tends to be needed for the larger and complicated rather than the smaller, simple projects. In many cases, developers act as their own project manager with 'in-house' staff or employ one of the professional advisers to exercise the management function. Typically, a project manager will receive a fee representing 2–3% of the final building cost, depending on the extent of the role and the complexity of the scheme. However, the developer may appoint a project manager on an incentive basis linked to whether the final cost is within budget. A development company may be asked to take on the role of project management on the basis of a fee, either fixed or related to the profit of the development, by an owner-occupier or property investment company for instance. Project management in this context has a much wider definition to include the management of the entire development process.

Project management is an occupation in its own right. A project manager may be an architect, quantity surveyor, a valuer/agent or, more usually, have a building/contracting background. The project manager should be appointed at an early stage as part of the development team to be able to advise the developer along with the quantity surveyor on the type of building contract applicable to the particular development, and to be involved in the development brief and the design discussions. In addition, the project manager should be able to advise the developer on the selection of the professional team, particularly those who have previously worked with the project manager. The professional team should complement each other and work well together. The project manager's role is to act as the client's representative when co-ordinating the professional team and liaising with the contractor. The project manager should be solely concerned with the overall management of the project and not be in any way involved in the work of actually carrying out any part of the project. The project manager needs plenty of common sense and administrative ability and a general working knowledge of the construction of buildings.

The management objectives must be clearly defined in consultation with the project manager and made known to all those working in the project team. Basically, the objectives are to ensure that the finished project will be suitable for the purpose for which it is intended, that it is built to satisfactory standards, that completion takes place on time and that the whole job is carried out within the approved financial budget. The project manager is very often responsible for appointing the professional team on behalf of the developer and, in this regard, will agree the fees, letters of appointment and deeds of collateral warranty under the guidance of the developer. The developer should ensure that the project manager is supplied with copies of all the funding documentation entered into with any financier of the scheme. The documentation will include plans and specifications agreed with the financier and the project manager should check these are complied with throughout the project. If alterations are necessary then the approval from the financier will be formally required. The project manager is also responsible for ensuring that arrangements for the disposal of the building, either the letting or the sale, are also carried out efficiently and satisfactorily.

It is essential to examine the role of the project manager through the pre-contract and contract stages. The following is based on a traditional JCT contract.

6.6.1 Pre-contract preparations

The project manager should check that the developer has the necessary legal title on the site, whether it is freehold or leasehold, and that vacant possession of the whole site is available immediately. All restrictions on the site should be carefully checked (e.g. underground services, easements and rights of light or support), and compared with the proposed scheme so that the building work will in no way interfere with them. The project manager will arrange (if not previously carried out by the developer) all the necessary ground investigations, structural surveys and site surveys, and communicate the results to the rest of the professional team. It is important that all the site boundaries are clearly defined, and that a schedule of condition of the boundary fences, adjoining roads and footpaths, etc., is prepared. It may be necessary to negotiate 'rights of light' or 'party wall' agreements with adjoining landowners/occupiers.

The architect is responsible for ensuring that all the necessary statutory approvals have been obtained, such as planning permission and building regulations. The fire officer should be consulted early in the design process and the architect should ensure that the design is in accordance with all relevant legislation. The architect is responsible for assuring the project manager that all necessary statutory consents have been obtained. It is most important for the project manager to obtain unqualified assurances on these matters because, in practice, many expensive delays are caused as a result of one or other of the statutory consents not being obtained before the contract starts. Sometimes there are circumstances which might persuade the project manager to allow a contract to start before all the statutory consents have been obtained, but in so doing the project manager and the developer must realize the risk that is being taken.

6.6.2   Preparing the contract documents

Arguably the project manager's most important job is to ensure that the contract is not allowed to start on the basis of inadequate documents. Incomplete drawings are probably the most common cause of contract delays and cost increases. If a contract is started before all the drawings are 100% complete and the architect is unable to provide all the drawings in time to meet the contractor's required schedules, the consequences can be serious indeed. The project manager needs to be absolutely satisfied with the availability of the drawings by the architect and that sufficient staff resources within the architect's firm are in place. If a contract is started before the drawings are 100% complete, which is usually normal, a detailed schedule must be obtained from the architect, showing exactly when the outstanding drawings will be delivered to the contractor. Before the building contract is actually placed, the architect must obtain from the contractor a written statement confirming that (provided the drawings are supplied in accordance with the architect's schedule) there will be no claims for delays due to lack of drawings. The importance of getting this matter right at the outset cannot be overemphasized.

Project managers must also be satisfied that the Bill of Quantities is as complete and accurate as possible. The quantity surveyor will measure the quantities off the architect's drawings, so again this stresses the need for their accuracy. Some items in the Bill of Quantities may be described under the headings of 'prime cost' (meaning actual cost) or 'provisional sums'. 'Prime cost' items usually cover materials or goods which usually cannot be precisely defined. 'Provisional sums' items cover elements of the work which it is not possible to detail properly and evaluate at the time the contract is entered into. The contractor will be asked to allocate a sum of money against such items. The project manager must understand why the prime cost and provisional sums items have been included in the Bill of Quantities, in other words, be satisfied that it is not possible to make the necessary detailed provision at the outset. Quantity surveyors should be closely questioned to ensure that they have received adequate information from the architect to enable them to prepare their Bills with complete confidence in their accuracy.

If a pre-letting has been achieved then it is important to include any specific requirements of the tenant within the contract document. Furthermore, such requirement should be clearly referred to as the 'tenant's specification' so there is no doubt. There may be a situation where the tenant subsequently alters their specification which delays the main contract on the 'developer's specification'. If a claim is subsequently made by the contractor then it can be apportioned to the tenant for payment.

6.6.3   Appointing the contractor

If it is proposed to invite competitive bids from a selected list of contractors, the project manager should agree with the architect and the developer the names of the contractors who are going to be invited to submit tenders. When the competitive tenders have been received and evaluated, the job is normally awarded to the lowest tenderer. The quantity surveyor will compare each tender against their own priced Bill of Quantities. The project manager decides whether a performance guarantee bond has to be obtained by the contractor. Once satisfied on all matters, the project manager then authorizes the placing of the building contract. All the contract documents should be ready, so that the contract may be signed before work actually starts on site, although in practice work often starts before the documents are signed, on the basis of a letter of intent, but this should be avoided.

The project manager will discuss with the architect reasons for wishing to appoint any nominated subcontractors and, if appropriate, then authorize their appointment.

6.6.4   Site supervision

The project manager should be continually satisfied about the arrangements made by the architect for site supervision during construction. The size and complexity of the scheme may merit the appointment of full time site supervisor, such as a clerk of works or a resident engineer or indeed a resident architect. The architect should also arrange for progress photographs to be taken periodically on site, so that a clear visual record of the state of the contract at any particular time is always available to supplement the architect's own reports on the progress generally.

6.6.5   Construction period

Once the contractor has taken possession of the site, the project manager is responsible for ensuring that the works are carried out on schedule and that the overall cost is kept within the budget. To enable the project manager to carry out his duties effectively, regular meetings of the project team are organized. The frequency and composition of the meetings depend upon the size of the particular job and might vary at different stages of the job. The project management meetings are often arranged on a monthly or fortnightly basis. The project manager, the architect and the quantity surveyor form the nucleus of the project management team. If the project manager is also controlling the letting or sale of the project, then the surveyor/valuer/agent is normally a member of the team, particularly in those cases where the purchasers or tenants might wish to have special works carried out. If the scheme is being financed externally, then the fund or bank's representative or appointed advisers will also attend the project meeting to fulfil their monitoring role. The contractor may be invited to attend the part of the project management meeting at which the progress on site is discussed. Very often, though, the project manager attends separate site meetings with the architect to be kept informed of building progress. All project management meetings should be carefully minuted.

Typically, at the beginning of a project meeting the minutes of the previous meeting are considered and any matters arising dealt with. The architect presents a report on the progress of the work, indicating what parts of the work are ahead of, or behind, schedule and whether the overall progress of the job appears to be satisfactory. The architect should indicate any difficulties which have arisen and state at every meeting whether or not the contractor is in any way delayed as a result of lack of information. The architect should also report as to whether any variation orders or architect's instructions have been given to the contractor, and if so, their likely effect on the progress of the work. The project manager will learn independently from the contractor or through attendance at site meetings whether the contractor is being delayed by lack of information or materials/labour.

The quantity surveyor then presents a report on the financial situation, indicating whether or not the work of measurement on site is well up to building progress and whether any variation order or architect's instructions have been given which affect the cost of the job. The quantity surveyor should also indicate the position with regard to prime cost and provisional sums and present an overall summary as to how the cost of the job so far compares with the contract sum. The quantity surveyor should also indicate any factors which might increase or decrease the cost of a job at a future date.

When appropriate, the surveyor reports on the progress with regard to the disposal of the property and on any requests for special or extra work received from prospective purchasers or tenants. Then the practicability and advisability of carrying out those special works are discussed. Ideally purchasers or tenants should take over the completed building in accordance with the original design and specification, carrying out required special works at their own expense once the building has been handed over to them. However, it is not always possible to insist on such an arrangement, as it may be necessary to carry out such works in order to secure the letting or sale.

Then the project manager summarizes the overall financial situation, particularly with regard to payments to the contractor, compares them with the budget, checks on dates of handover and compares the estimated date for the receipt of income or capital payments with the budgetary expectation. These are matters of vital importance to the developer's cash-flow. If at any stage it appears that the project is running behind schedule, then methods of speeding up the work to recover the position need to be considered, together with the implications for cost. Usually there is a liquidated damages provision in the building contract and the question of its enforcement has also to be considered. In practice, liquidated damages are often inadequate to compensate the developer for losses incurred as a result of the delays, because if the true cost is written into the contract documents at the time of the invitation of tenders, contractors would increase their tender prices out of all proportion in order to safeguard themselves against a the risk of a heavy liquidated damages claim, which might in fact never be made.

This summary of project management arrangements where a JCT contract is used is intended to illustrate the basic principles involved, which apply generally to other procurement methods. The project team may be much larger and a management contractor or a construction manager may be part of the team on building schemes of a highly complicated nature. Accordingly, the project management is that much more intricate. On the other hand, where design and build contract methods are used, the role of project management is simplified accordingly, and may even be taken over by the quantity surveyor. In the case of a lump-sum, fixed-price design and build contract, project managers are essentially concerned with quality control and progress. They may inspect the buildings during construction or arrange for a professional adviser to do so. Periodic meetings with the contractor to discuss building progress and the achievement of the handover dates should enable them to fulfil their role.

6.6.6 Handover

A short time before the date for completion and handover of the building from the contractor to the developer, the architect prepares a 'snagging' list, indicating all the minor defects that must be remedied before handover takes place. At that time, it is useful for the developer's surveyor and the representative of the intending occupier (if known) to accompany the architect to ensure that they both are satisfied with the snagging list which has been prepared. At the outset of the contract, the project manager will have confirmed that the building works are adequately protected by the contractor's own insurance arrangements. The contractor's insurance will no longer protect the building once it has been handed over, so one of the most important things is for the project manager to ensure that the developer has adequate insurance cover from handover until such time as the insurance cover provided by the occupier takes effect.

If the development has been pre-let or pre-sold to an owner-occupier, then the occupiers and/or their contractors may wish to have access before formal handover by the main contractor working for the developer. From the developer's point of view, this situation should be avoided at all costs, unless it is necessary to secure the deal. If an occupier wishes to gain early access to attend to fitting-out works, then arrangements should be documented clearly in the contract. Ideally an occupier's special requirements should be incorporated at an early stage into the design process or if the occupier is secured after the building contract has started, then the occupier should only be allowed access after practical completion of the building. It is important to include the developer's base specification for the scheme into any documentation, so that any changes which lead to an increase in cost or delays can be attributed to one party or the other. If there is an overlap between main contractor and fitting-out contractor, the project manager should ensure that the occupier arranges adequate insurance.

Difficulties arise when the work of the fitting-out contractor affects the work of the main contractor. The project manager together with the architect need to attribute and resolve problems quickly.

The quantity surveyor should then be asked when any outstanding remeasurement work will be completed and be in a position to agree the final account with the contractor, so that the architect may issue a final certificate. The JCT contract will have provided for a certain percentage of the total cost to be retained by the building owner until the end of the defects liability period, often 6 months from the date of practical completion (12 months in the case of any electrical and mechanical element of the contract). Special maintenance periods may be agreed for particular parts of the work (e.g. landscaping). The contractor is responsible for remedying any defects (other than design) which have occurred during the defects liability period, provided of course that they have not been caused by the occupier. It is most important that the buildings should be carefully inspected at the end of the liability period, because if there are any obvious defects at that time which the architect does not identify, it may well be assumed that the architect was prepared to accept the building subject to those defects.

The importance of inspecting the site and its immediate environs on the handover date should not be overlooked. If during the building contract any damage has been caused to adjoining property damage to boundary walls and fences is not unusual, then the contractor must be required to remedy it. Inspection of the roads, footpaths, kerbs, grass verges, etc., immediately adjoining the site is carried out to see that the contractor remedies any damage, otherwise the highway authority might subsequently ask the developer to bear the cost of any remedial works.

The architect should produce as 'built drawings' a building manual and maintenance schedule to assist the occupier by giving a comprehensive schedule and description of all components (taps, locks, fastenings, sanitary ware, etc.) which might need replacing at some future date, together with recommendations for regular maintenance work to preserve the building fabric.

Where an occupier is not taking possession immediately, then the developer is responsible for a vacant building, and a programme of regular cleaning and maintenance should be instigated. Whatever physical arrangements reasonable and necessary to protect the property against vandalism should be made. An obvious example is in the case of a shopping development, where any unlet shop units will have a neat hoarding put across the frontage immediately before the handover date. Consideration is often given to the advisability of employing security guards or patrols. Adequate insurance cover should be in place to give protection against fire and loss due to the damage of property. Public liability insurance should also be arranged to protect against claims from injured third parties.

6.6.7   Monitoring construction progress and costs

The project manager's objective is to produce the building on time and within budget for the developer client. Therefore, it is important to examine how the project manager reports to the developer on construction progress and cost. Any delays in completion or increase in costs will affect the profitability of the development, therefore it is essential that a developer is kept regularly informed on progress and cost. The developer will need regularly to update the cash-flow appraisal prepared at the initial evaluation stage (see Chapter 3) to assess profit.

Every project manager will have their own method of reporting but it is important to agree with the developer at the outset the information required. The best method of reporting uses charts and graphs to compare actual progress and cost against the original estimates. The starting point should be the appraisal used at the site acquisition stage. Actual costs and progress are compared against the estimates made at the time of acquisition. This means the developer can at a glance clearly identify changes in costs and progress, instead of reading through pages of written information to obtain the same information. The charts usually will have written comments on them providing reasons why costs have increased/decreased or why progress on site is behind schedule. Once the developer has assimilated the information presented in the charts and graphs further questions can be asked of the project manager as to the reasons behind identified increases in cost or delays in process. In particular it should be clearly understood who is able to authorize the expenditure of money. Every member of the project team must know whether they are able to spend money and, if so, what authorities they must obtain.

Typical reporting methods include the following:

(a) Bar chart

The bar chart is a calendar showing the development programme, either in weeks or months. The programme is divided into various tasks and the period during which each of these is to be carried out is shown on the chart. An example of such a chart is shown in Figure 6.1. You will notice that the chart includes pre-contract activities as well as the contract programme, which are equally important to monitor as any delay will have an impact on the start of construction. This demonstrates how important it is for the project manager to be involved in the development process from the beginning. The chart clearly indicates when each task is to start and when it is to finish. It shows how the various tasks overlap and the work that should be in hand at any particular time. From time to time the programme, and therefore the bar chart, may need to be changed but by comparing what has actually been achieved with what the chart shows gives the developer and project manager a simple yet instant test of progress.

The bar chart can also be used to indicate when information or decisions are needed by the project manager from the developer and by the contractor. This is very important as lack of information or instructions is one of the principal causes of delay. The bar chart clearly demonstrates that any delay in one activity can affect the whole programme. Once a delay has been identified it is important for the project manager to advise the developer what effect it will have on the overall programme and how time can be made up in other activities. It is important that the project manager issues the bar chart to the whole of the professional team, so that each of them can identify the target dates they have to work to.

The bar chart may be substituted once the contractor is on site, with the contractor's bar chart which identifies the timescale for each trade involved on site. It is important that the project manager receives the contractor's bar chart regularly so that the overall bar chart can be updated and amended as necessary. The developer may not need to know the progress of each individual trade on site and it is sufficient to break down the contractor's programme, into substructure, superstructure, finishes and external works. However, the project manager must be able to produce the contractor's bar chart at any time, as in some cases the developer will need to know the detailed programme. For instance, the developer may need to know when the area in the building identified as a show suite is ready (see Chapter 8).

Another method of monitoring progress and highlighting the importance of providing information and decisions is to prepare a chronological timetable of events. However, the bar chart is the most instant way of comparing progress against original estimates.

(b) Cash-flow table and graph

A cash-flow table and graph should be prepared by the project manager, examples of which are shown in Figures 6.2 and 6.3. The purpose of the cash-flow table in Figure 6.2 is to estimate the developer's flow of cash payments throughout the development period. The developer can use this to prepare a cash-flow appraisal, which can be regularly updated throughout the development. The importance of the cash-flow has already been discussed in Chapter 3 on Evaluation. The table can be represented as a graph as shown in Figure 6.3. The combination of the table and graph can provide another means of checking progress by comparing actual payments with estimated payments. However, it is less effective at measuring progress than either the bar chart or a development timetable. Estimates of cash-flow often have to be revised and there is a danger that they do not really highlight problems of delay until the last months of the contract.

(c) Financial report

The project manager's financial report may typically look like the example shown in Figure 6.4. It is based on the quantity surveyor's cost reports and payments already made to the contractor as certified by the architect. It enables the developer to identify variations in costs throughout the contract.

The project manager should advise the developer of the reason for the cost variation. Any variation in cost from the original contract value may be due to a claim from the contractor, architect's instructions to the contractor and variations required by the developer. We have already exam-

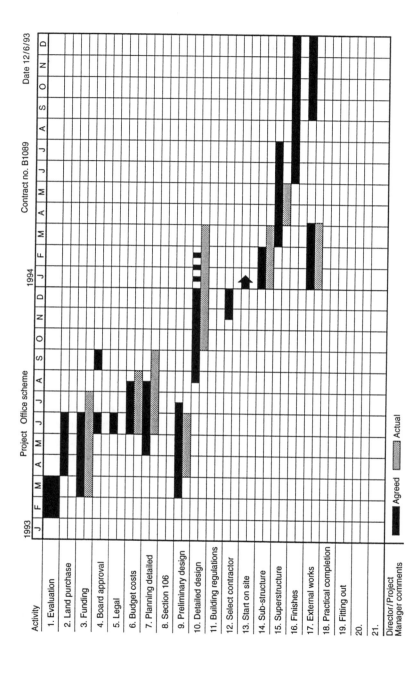

Figure 6.1 Overall development programme – bar chart.

Figure 6.2 Cash-flow: fees/construction.

Project Office scheme · Contract no. B1089 · Date 12/6/93

Fees (£000's)	Budget £	Total to date	1993 J	F	M	A	M	J	J	A	S	O	N	D	1994 J	F	M	A	M	J	J	A	S	O	N
Architect	225	168						15		15			15		10				6						
Structural eng.	90	45						12		11			11		6				6						
Quality surveyor	90	45						11		12			11		6				6						
M & E engineer	68	34						17																	
Project manager	90	40								20			22		17				10						
Acoustic																								
Landscape																								
Right of light																								
Party wall																								
Site surveys	3	3																							
Ground surveys	8	15								9															
Planning	3	3																							
Building regulations	22	13																							
Others																								
Demolition																								
Enabling																								
Main contract	4500	765						698	799	461	474	528	318	225	71	108									
Stat. authorities																				138				
Fitting out																								
Budget total	5099	1349						459	405	426	540	585	597	360	215				28		135				
Actual total	5199	1131						753	799	528	474	528	377	225	110	108			28		138				
Building/Project Manager comments																								

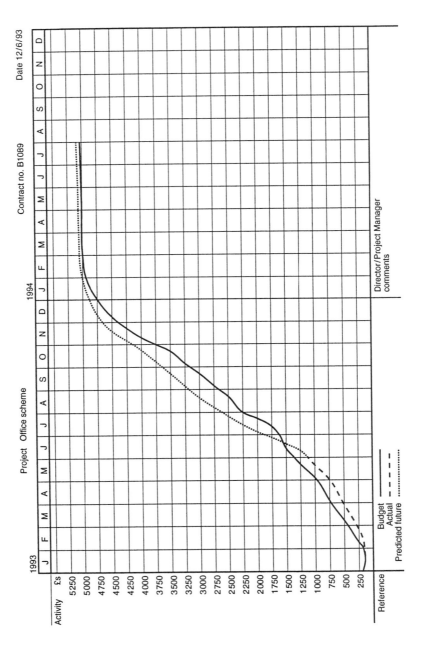

Figure 6.3 Development costs: building costs graph.

Project Office Scheme   Contract no. B1089   Date 12/6/93

#		Board app. Date 15/9/92	Revised report Date 10/5/93	+(-)	Revised report Date 10/5/93	+(-)	Revised report Date	+(-)	Director/Project Manager comments
1.	Site start	8/1/93	8/1/93		8/1/93				
2.	P/completion	21/12/93	28/1/94	-one month	28/1/94	-one month			Agreed extension of time see memo
3.	Net lettable area	38250	38250		38250				Client to measure in July
4.	Building contract value	4500000							
5.	Demolitions								For cost breakdown see QS report no. 6
6.	Enabling works								
7.	Sub-structure			+53034		+53034			Increased cost of piling due to unforeseen ground conditions
8.	M & E services					-6740			Saving due to design change
9.	Finishes		3953034		3946294				
10.	External works								
11.	Preliminaries								
12.	Inflation								
13.	Contingencies	100000	100000		100000				£15456 of contingency not expended
14.	Statutory services								
15.	Tenant works								
16.	Claims (unsettled)								
17.	Instr. (not priced)								
18.	Pending instructions								
19.	Others agreed extension of time		23250	+23250	23250	+23250			Delay caused by late receipt of drawings from structural engineers
20.									Clients variation to finishes budget needs to be finalized
21.	Clients variation				15000	+15000			
22.									
23.									
24.									
25.									
26.									
27.									
28.									
	Total	£4500000	£4576284	+£76284	£4584544	-£84544			

Figure 6.4 Financial report: building costs.

ined the circumstances under which a contractor can make a claim for additional costs. Claims may be based on the inadequacy of the drawings and/or the Bill of Quantities. In addition, they may be based on delays in the architect issuing drawings or instructions. The project manager must therefore ensure that claims and variations are kept to a minimum if costs are to be kept in budget. The project manager should carefully monitor the activities of the architect and ensure they keep to their drawing schedule. The project manager together with the quantity surveyor must advise the developer of the cost of any variation proposed to ensure the developer is aware of the implications. No revision should be made without justification. The developer must know why cost estimates have to be revised. The project manager must maintain a tireless scrutiny of costs and question any decision thoroughly which has a cost implication.

(d) Checklist

It is a good discipline for the project manager to prepare a checklist of the main activities throughout the development. An example of such a checklist is shown in Figure 6.5. The checklist will define the main activities applicable to the particular development, some of which will require approval by the developer. It should also highlight information required by the project manager from the developer. It is essential for the project manager to identify regularly the decisions needed by the developer and by what date. Developers should be aware if the progress of a development is being held up because a decision is required of them, and the implications of any delay in the decision. The developer/project manager relationship is a two way one and both should ensure the other is kept fully informed at all times.

The above mentioned methods of reporting are typical but every project manager will have their own method of reporting. Whilst regular reporting on progress and cost is a way of keeping the developer informed, it also provides the project manager with an essential tool. If the developer insists on regular reporting in the manner shown above then the project manager will know whether the aims are being achieved. It will bring into sharp focus the targets that need to be achieved and the problems that need to be tackled.

The project management of a development through the construction process, whether carried out by the developer or through the appointment of a project manager (or other professional) is all about teamwork and motivating the participants to work together. Problems must be

Project Office scheme   **Contract no.** B1089   **Date** 12/6/93

	Month	Approved by	May 1993 Project Mgr.	May 1993 Client	June 1993 Project Mgr.	June 1993 Client			Director/Project Manager comments
1.	Client brief - agreed		✓	✓	✓	✓			
2.	Select/appoint architect		✓	✓	✓	✓			
3.	Select/appoint QS		✓	✓	✓	✓			
4.	Select/appoint str. engr.		✓	✓	✓	✓			
5.	Select/appoint M & E		×	×	×	×			All agreed except one item
6.	PM appointment		✓	✓	✓	✓			
7.	Deeds of coll. warranty		×	×	×	×			All agreed except architects
8.	Letters of intent (specify)								
9.	Development/feasibility appraisal			✓		✓			
10.	Site boundary/ownership agreed			✓		✓			
11.	Appoint right of light								
12.	Appoint party wall								
13.	English Heritage agreement								
14.	Summary of funding documentation			✓		✓			
15.	Funds/bank surveyors approval								
16.	Tenant requirements								
17.	Agreed contract programme		✓	✓	✓	✓			
18.	Planning drgs/application		✓	✓	✓	✓			
19.	Section 106 agreement								
20.	Freeze design/Stage D report		✓	✓	✓	✓			
21.	Bldg. regulations application		✓	✓	✓	✓			
22.	Certificate of readiness		✓	✓	✓	✓			
23.	Summary of ins. requirements		✓	✓	✓	✓			
24.	Summary of bldg. contract cond.		✓	✓	✓	✓			
25.	Appoint bldg. contractor		✓	✓	✓	✓			
26.	Agreement with stat. undertakers		×	×	×	×			
27.	Signed receipt for main manuals								
28.	Client decisions (major items)								
29.	Finishes board - urgent*		×	×	×	×			Meeting arranged the 16/6/93 to discuss
30.									
31.									
32.									
33.									
34.									
35.									
36.									
37.									
38.									
39.									
40.									

Figure 6.5 Main activities: checklist.

sorted out before members of the team start to blame each other and become entrenched. The project manager must be able to anticipate delays by ensuring constant communications with the professional design team and the contractors. Contracts and paper communications should not be relied on: there is no substitute for personal contact. Overall project managers should fulfil their role efficiently, constantly bearing in mind both cost and time. They need the ability to lead and motivate the professional team and the contractor. Again, it shows that the property development process is all about the interaction between people.

Executive Summary

The developer's aim during the construction process is to produce a good quality building on time and within budget. There are three main types of building contract available: traditional JCT Contract, design and build, and management contracting, although there are many variations on each depending on the exact contractual arrangements and the role of the professional team. Each contract has its main advantages: the traditional contract for its flexibility; the design and build for cost control; and management contracting for speed. Due to widespread disillusionment by developers with the traditional contract the design and build contract is becoming increasingly favoured as the contractor is responsible for design and construction, avoiding the problem of cost increases due to the late production of drawings by an architect. To overcome the problem of quality control with the design and build contract the appointed contractor is taking over the appointment of the developer's chosen architect through a novated contract arrangement. Whichever method is used the success of the building contract, in terms of achieving the aims above, relies on good control, leadership, and firm early decision making by the developer and the project manager. The professional team and the contractor should be motivated towards the same goal, resolving any conflicts and problems before they arise.

REFERENCES

Masterman, J.W.E. (1992) *An Introduction to Building Procurement Systems*, E & FN Spon, London.

CHAPTER 6 CASE STUDY – USE OF A DESIGN AND BUILD CONTRACT ON A RETAIL PARK

This case study provides a good example of the successful use of a design and build contract, although in a varied form, on a completed edge of town retail park. The developer was successful in both pre-letting and forward-funding the scheme, and so by the time a decision had to be made on the form of building contract the only remaining risk to the developer was the building cost and time. The developer had never used the design and build procurement method before but was thoroughly disillusioned with the traditional JCT contract, due to the experience of excessive claims from contractors. We will briefly examine how the standard design and build contract was varied and administered on this scheme, focusing on the construction phase of the development.

BACKGROUND

The developer specializes in retail development, both town centre and out-of-town schemes, and they have carried out some office development. The development, which is the subject of this case study, was a green field site on the edge of a town on an 'A' road, adjacent to an existing major food superstore. The developer, having obtained detailed planning consent for a scheme comprising 70 000 sq ft (6503 sq m) of non-food retail and a 13 000 sq ft (1208 sq m) food supermarket, secured pre-lets on the entire scheme which is usual with such out-of-town retail schemes. The tenants include a supermarket, a garden centre and some of the smaller DIY and home furnishing retail multiples.

With such pre-lets in place the developer was able to secure forward-funding with a fund. The development is currently being extended to include outlets for two major fast food retailers.

THE CONTRACT AND ROLE OF THE PROFESSIONAL TEAM

The developer had appointed the architects, the quantity surveyors, Silk and Frazier, and a structural engineer, from the beginning on the basis of adopting the traditional JCT route. However, at the point where a decision had to reached on the type of building contract it was collectively decided that the design and build route would be adopted instead, due to recent bad experiences by all concerned (for all the reasons we have previously examined in this chapter).

However, in order to maintain some control over the quality of the design and construction the developer wanted the architect, the quantity surveyor and the engineer to remain involved in the scheme throughout the construction process. So it was decided that a variation on the standard design and build contract would be adopted. The developer would novate the contracts (letters of appointment) of both the architect and engineer to the contractor. In other words the contractor would become the client of both professionals on identical terms to those contained in their letters of appointment with the developer. The contractor would then be responsible for carrying through the detailed design and construction of the scheme with the appointed professionals. In addition, the developer also appointed the same architect directly to be responsible for quality control. As the architect was responsible to two clients and a potential conflict of interest existed, it was decided that one partner in the practice would work for the developer and another partner would work for the contractor.

Once the design and build route had been decided on the architect was instructed to prepare the performance specification incorporating the various tenants' specifications in great detail, including a specification of all the materials to be used. The tenants would be responsible for their own fitting out but the developer's appointed contractor would incorporate such items as fire protection and toilet facilities within the main contract. It is usual where the developer bears the cost of such tenant's facilities that they become landlord's fixtures and fittings under the terms of the lease, and are taken account of for rent review purposes.

At first, the performance specification was sent to a major contractor. A price was submitted to the developer which was considered a little on the high side. So another contractor was invited to tender; however, although they came up with a lower price, the fund would not consider them due to concern about their financial standing. So, the first contractor was appointed as contractor on a negotiated 'JCT 1981 Design and Build Contract With Contractor's Design', but on the basis of novated contracts with both the developer's architect and engineer. The contract price equated to £27.14 p.s.f. (£292.13 p.s.m.) and a programme of 6 months was written into the contract, although it was generally accepted by all parties concerned that this was rather tight. The supermarket unit became a separate contract between the retailer and the contractor as the retailer had acquired a long leasehold interest in their site from the fund.

At the time the contract was being negotiated (late 1993) tender market conditions favoured the developer's negotiating position as competition was fierce for the few building contracts available. This enabled the developer to shift most of the risk of increases in costs and delays onto the contractor. It is important to remember that a developer will always aim to minimize risk at every opportunity during the development process whilst maintaining a reasonable profit level. Under the terms agreed on this contract the contractor would only be able to claim an extension of time if any delay was caused by the developer's variations to the contract. The standard clauses allowing the contractor to claim in the event of delays caused by bad weather or changes in legislation were deleted from the contract. In addition it was also agreed that any variations made by the tenants would only be accommodated by the contractor if they did not cause any delay on the main contract relating to the developer's works.

The entire pre-contract period took about 6 months. A lot of this time was taken up by the collateral warranties which had to agreed on behalf of the fund with the professional team and the contractor (together with subcontractors). As we have already discussed in this chapter, collateral warranties involve the developer in endless negotiations with the solicitors acting for the fund and the various professionals and contractors.

Silk and Frazier, the quantity surveyors, who had a long standing working relationship with the developer, were appointed as the 'employer's agent' (developer's representative) for the purposes of the contract. Usually, this would involve the quantity surveyor as both the contract administrator (in place of the architect who traditionally performs this role under a traditional JCT contract) and project manager. In fulfilling such a role the quantity surveyor would then be responsible for appointing the professional team and agreeing their deeds of collateral warranty; chairing and minuting all project meetings; and preparing all interim valuations and the final account. However, on this particular scheme what in effect happened is that the architect chaired and minuted the meetings and the quantity surveyor undertook a normal role. The fund were represented at all the project meetings by their own in-house structural engineer, quantity surveyor and building surveyor.

CONCLUSION

The contract took just over 7 months and finished in April 1994. This was 6 weeks late. However, as the developer acknowledged that the original programme had been tight, a compromise was made over the payment of liquidated damages by the contractor. The contractor was due to pay £25 000 a week under the terms of the contract but this was varied to £1000 for the first 4 weeks and £25 000 for the remaining 2 weeks. The final account is yet to be agreed (November 1994) and the contractor has presented a claim for £100 000. It is believed that the final outcome will favour the developer and any claims settled will be very minimal. Overall, the developer is generally pleased with the financial outcome of the contract as it would appear that, subject to negotiations with the contractor, the scheme is on budget.

In addition, the developer is reasonably happy with the quality of the scheme and will certainly be using design and build again on future projects. Keeping the two roles of the architect separate has not been entirely easy and is probably not the ideal way of arranging such appointments.

Currently, this variation of the design and build contract, with novated professional appointments and with quantity surveyors acting as the developer's representative, is becoming more widespread. Silk and Frazier estimate that 60% of their current appointments are design and build contracts. Whether this trend will continue will depend on the tender market as currently contractors are having to accept a considerable share of the risk in order to obtain work from developers. Some developers are now using design and build on all their developments.

7

MARKET RESEARCH

Jayne Cox, Property Market Analysis

7.1 INTRODUCTION

The third edition of this book (1991) contained only a brief introduction to the very fast growing area of property market research and it urged:

> greater emphasis needs to be given to the research aspects of the property market by those designing courses for students wishing to enter that market, either as chartered surveyors or as specialist market analysts.

This chapter provides a more extensive survey of property market research and the ways in which market information and analysis aid the development process. It takes further the distinction made in the third edition between information and analysis, and provides an introduction to the emerging specialisms of property market forecasting and portfolio analysis. The chapter is intended as an introduction to property market research for those who will become users and purchasers of research services as well as those who are starting out as analysts.

7.2 SUPPLIERS AND USERS

7.2.1 Suppliers

Property market research grew very fast during the 1980s and it has quickly become a recognized area of expertise for property professionals. The Society of Property Researchers, which began life in the mid-1980s, now boasts some 320 members. Property research, meanwhile, has been recognized by the Royal Institution of Chartered Surveyors as a qualifying element for the Assessment of Property Competence. Most large commercial estate agency practices now offer research services and the major property investing financial institutions almost all have in-house research teams. A number of specialist information providers and independent consultancies are also now well established, e.g. Applied Property Research (APR), Property Intelligence, Investment Property Databank (IPD) and Property Market Analysis (PMA). Several of the largest management consultancies also offer property analysis and consultancy services. These organizations provide research and information principally to commercial clients.

In addition to commercial property research, the discipline boasts a growing academic literature (e.g. see *Journal of Property Research*). Much of this new literature is being generated by university departments, such as those at Reading and Aberdeen. However, as well as conducting 'basic' research (trying to understand the fundamentals of property markets), academics are increasingly commissioned to do research by external clients, in competition with private sector consultancies and agency departments. Collaboration between academic researchers and commercial market researchers is lamentably rare, however. In part this reflects the difficulty of obtaining funding to publish research: the results of most research projects are commercially sensitive and, understandably, clients are wary of wider publication of what is relatively expensive information and advice. A notable exception, however, is the recent Royal Institution of Chartered Surveyors sponsored report on *Understanding the Property Cycle*, authored jointly by teams from Aberdeen University and IPD (Royal Institution of Chartered Surveyors, 1994a).

7.2.2 Users

The principal users of research are large development companies and investing institutions, together with their agency teams. The public sector – local authorities, Urban Development Corporations and central government – also commission external property market research from time to time.

A distinction between suppliers and users, however, is not always strictly appropriate. Many organizations involved in property market research will be both suppliers of research to external clients and users of other people's. Research teams in the large institutions, for example, may commission outside consultants to provide additional information or analysis on specific projects while they will also have their own internal 'clients' to whom they provide advice and filter external market research. An example of how an organization uses and provides research is illustrated by the Prudential, which has one of the largest in-house property research teams amongst the institutions (Figure 7.1).

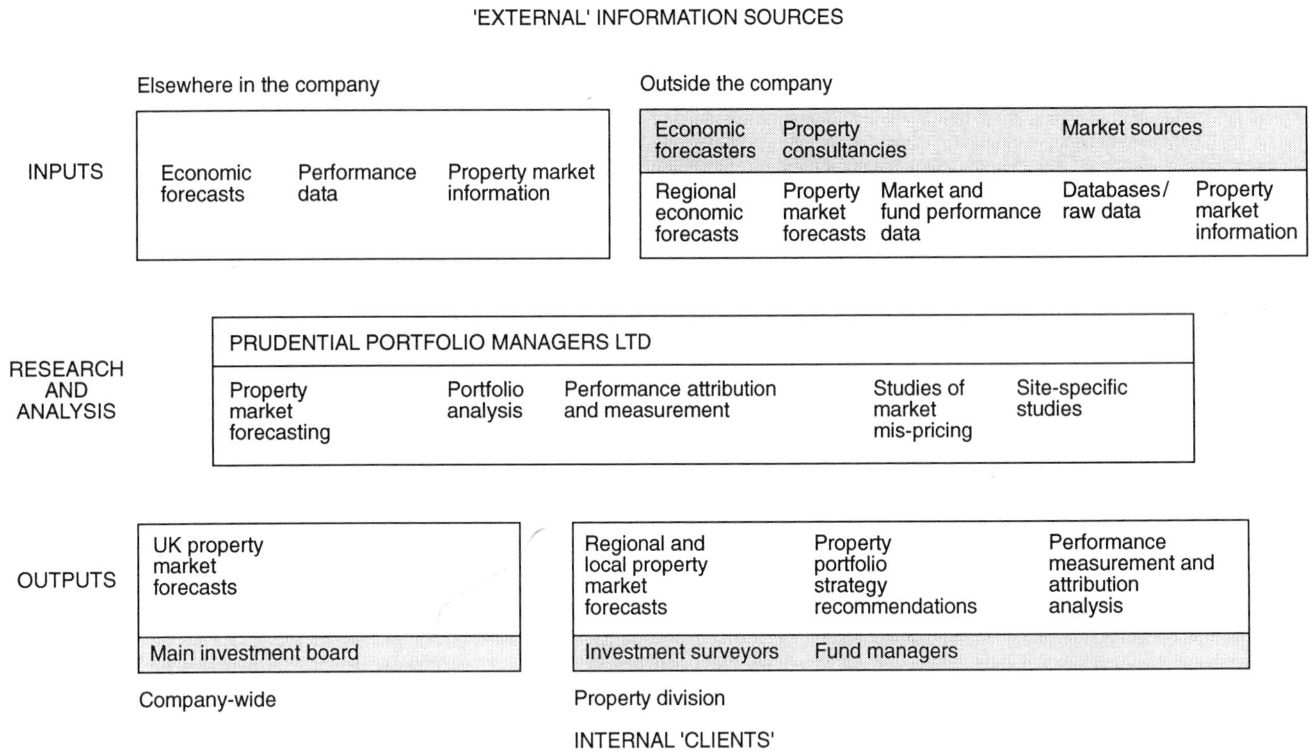

'EXTERNAL' INFORMATION SOURCES

Elsewhere in the company | Outside the company

INPUTS

Elsewhere in the company:
- Economic forecasts
- Performance data
- Property market information

Outside the company:

Economic forecasters	Property consultancies		Market sources	
Regional economic forecasts	Property market forecasts	Market and fund performance data	Databases/ raw data	Property market information
	Studies of market mis-pricing		Site-specific studies	

RESEARCH AND ANALYSIS

PRUDENTIAL PORTFOLIO MANAGERS LTD

- Property market forecasting
- Portfolio analysis
- Performance attribution and measurement

OUTPUTS

UK property market forecasts	Regional and local property market forecasts	Property portfolio strategy recommendations	Performance measurement and attribution analysis
Main investment board	Investment surveyors	Fund managers	

Property division

Company-wide

INTERNAL 'CLIENTS'

Figure 7.1 Use and supply of research – Prudential Portfolio Managers. Source: P. McNamara, Prudential Portfolio Managers Ltd.

7.3   THE USE OF PROPERTY MARKET RESEARCH

The use of property market research is well established:

1. To inform decisions on individual developments or investments; often to test and refine the initial hunches of the developer or investor.
2. To help in the formulation of long-term development and/or investment strategies.
3. Related to the last point, to analyse and predict the performance of property investment portfolios.

Each of these uses will bring together a mix of different research skills and approaches. These are discussed further in Section 7.4.

In addition to these established uses, property market research is likely be used more widely by valuation surveyors in future, following the recommendations of the *Mallinson Report* on commercial property valuations (Royal Institution of Chartered Surveyors, 1994b). As the report concedes, valuation has often in the past been seen as a somewhat mystical discipline. In recognition of the criticisms made by some valuation clients in the course of consultation for the report, Mallinson urges greater 'transparency' in valuation practices, in particular, the provision of more extensive supporting evidence. The report recommends that valuers should be able to express informed opinion on:

1. The local economic factors affecting value.
2. How the valuation fits into 'trends' in the market.
3. The 'certainty' of the valuation given the uncertainties which surround the property 'product'.

While recognizing the importance of forecasting property markets – in so far as the future should be reflected in today's values – the report maintains that forecasting should continue to be seen as a separate discipline. Nonetheless, the expanding scope and use of property market research is such that surveyors entering the market in future will need to be conversant with the full range of research services available, if only to become sophisticated purchasers of them. The following section provides a guide to the main types of research.

7.4   TYPES OF RESEARCH

The third edition noted that:

As the market operators have become familiar with the use of information and consultancy, the distinction between the two has become clearer.

Moreover, since the last edition, a number of specialisms within the consultancy umbrella have become more widely established. This section discusses:

1. Information
2. Site-specific and strategic market analysis
3. Forecasting
4. Portfolio analysis

7.4.1 Information

We consider two different types of information:

a. Property market information
b. Supporting information

(a) Property market information

Official statistics on commercial property are sparse indeed. While some partial information is available on development activity, very little exists on transactions or prices. The Department of the Environment produces data on New Construction Orders (orders placed with contractors) which can be considered equivalent to development starts. A serious drawback of the data, however, is that they are presented as the value of orders. Since the Department of the Environment abandoned the collection of floorspace statistics in 1985 there has been no consistent source of information on the quantity of floorspace completed in any year or, indeed, the total stock of commercial buildings. Moreover, the New Orders statistics are available only at national, regional and county levels. Information on development activity at a more local level is rarely available on a consistent basis over long time periods, with some notable exceptions (e.g. in the City of London). Data on floorspace stock and development completions most usually have to be pieced together painstakingly from a number of disparate sources. As regards rents and investment activity, the Central Statistical Office (CSO) publishes data on pension fund investment; the Inland Revenue publishes a twice yearly *Property Market Report* which contains reports from district valuers, including some information on rents and capital values in selected towns. However, none of these official sources provides sufficient information for the surveyor or researcher to be able to glean a reasonable view of market activity, either at the national level or for individual markets.

900 Slough, 950–959 Buckingham Avenue, Slough. Ten industrial/warehouse units with a two-storey office content, varying between 25 and 30% of the unit floor areas. Their design is based on market research carried out by the developer Slough Estates plc and the quality of the office accommodation was found to be one of the important design factors from the user's viewpoint.

In the near absence of any official statistics on the subject, the property world has had to develop its own record of transactions and prices. The large agency practices now provide well regarded information on most aspects of the market, including building availability, transactions (take-up and rents) and performance (yields and capital values). The information is provided either in regularly produced bulletins or in occasional reports on sectors or topics of special interest. The sources are too many to list here but examples include the *King Sturge* series on availability of industrial space in England and Wales, *Healey & Baker PRIME* (rents for different property types in selected towns across the country) and the *Hillier Parker Average Yields* series.

Agents' most extensive coverage is almost certainly of the central London office market, a not at all surprising phenomenon when we consider that around one-fifth of institutional property investment (by value) is located there. However, rent and/or yield indices are provided for the UK standard regions by a number of agents (e.g. *Healey & Baker PRIME* and *Hillier Parker Average Yields*). Data on rents in selected towns are also available (e.g. *PRIME* and *JLW 50 Centres*). In addition to the large London agents, a number of firms in the largest provincial cities also produce market reports, e.g. the *Ryden Property Market Report* covering Scottish centres and reports by Lambert Smith Hampton on the office market in several provincial cities.

Surprisingly to outsiders, the property industry does not have an agreed 'industry standard' performance measure, similar to the FTSE 100 for example. The closest approximation is probably the IPD Index produced by Investment Property Databank, an organization sponsored by 15 firms of agents and holding a large database covering £40.5 billion of institutional funds in 1993. National rent, yield, capital value and returns indices are supported by further information on the performance of different types of property and different regions.

Two other property databases deserve mention. First, Applied Property Research (recently merged with Glenigan Property Research) holds extensive databases on office development in central London and the main regional cities; also details of planning applications for commercial uses across the country. Details of occupiers are available for certain locations and the company hopes to expand its geographical coverage in future. Secondly, the FOCUS on-line service, produced by Property Intelligence, provides direct computer access to a number of property databases. Information is drawn from specialist property and other journals across the UK and includes databases on reported property deals, ownership, rating and new development. Its Town FOCUS service also provides selected socio-economic and demographic indicators.

Property market information is currently delivered in printed reports or via telephone modem links which produce computer-based text or spreadsheet output. The next exciting step on the information front will be the introduction of Geographical Information Systems (GIS) into the storage, use and presentation of property data. Progress in this area is in its infancy and we look forward to reporting new developments in the next edition of this book.

(b) Supporting information

As property market analysis has evolved and become more sophisticated, it has drawn more and more on what the *Mallinson Report* calls 'supporting evidence'. As we shall see when we look at different types of analysis, however, this information is far from just 'supporting' and is often central to unravelling why the property market in general, or in a specific place, behaves in a particular manner.

For instance, if demand is analysed solely on the basis of deals completed on individual properties part of the picture may be missed: the number and type of completed deals, for example, is often restricted by the availability of suitable property. Further light can be shed on the underlying level and character of demand by looking at an area's demographic characteristics (especially for retail developments) or the structure and performance of its local economy, amongst other things. Official statistics of this type are available from the decennial Population Census and tri-annual Census of Employment; also from consultancies which provide additional analysis of Census data (e.g. CACI and URPI), local authorities and Training and Enterprise councils (TECs).

As well as indicators of local economic well-being, the analyst may also be required to advise on the overall state of the economy or particular occupier types, e.g. manufacturers or retailers. Many such macro-economic indicators are employed in property market analysis including, for example, statistics on Gross Domestic Product (GDP), retail sales and manufacturing output. These data are released by the CSO in a number of official publications, most importantly the *National Accounts* (or Blue Book), *Economic Trends* and *Financial Statistics*. Information on employment and unemployment is produced by the Employment Department. National employment statistics and selected data for regions, counties and local authorities is published in *Employment Gazette*; further geographical or industry detail is available (at a cost) from NOMIS at the University of Durham, where various Employment Department databases are held.

ıment statistics sometimes useful to property researchers can
ı, for example, *Regional Studies*, *Social Trends*, *The Family
ʾurvey*, *The Labour Force Survey*, *Transport Statistics* and *Housing
ın Statistics*.

ʾe survey of data sources (property and supporting) is by no
means exhaustive and the range of useful publications is increasing all
the time. Nonetheless, substantial gaps persist. All too often, researchers
and analysts are spending their (and client's) time piecing together information which might much more efficiently be provided via official
sources. A case in point is the almost total lack of information on commercial floorspace (although we understand the Department of the
Environment may soon address this issue). Moreover, the fact that property information has developed in a piecemeal fashion, led by competing
private-sector interests, means that it is often inconsistent between
sources (though usually consistent within any given source) and has
focused on areas which have attracted the most investment – central
London and retail. This latter aspect is not altogether bad so long as the
market remains focused in the same direction: the collapse of the central
London office market at the end of the 1980s, however, has shifted
investor and developer interests elsewhere and into locations and property types on which existing information is poor. As the next development
cycle gathers pace, some new and grand mistakes may well be made in
these poorly researched and understood markets.

7.4.2   Strategic and site-specific analysis

For some purposes it is often sufficient for the surveyor or researcher to
supply his/her client simply with information. As information has become
more widely available, however, clients have, quite rightly, come to expect
interpretation of the information in the light of specific questions about
their company or individual development plans. Here, we outline briefly
two different types of analysis which are commonly required:

a. Strategic analysis
b. Site-specific research.

(a) Strategic analysis

Strategic analysis is conducted either by a company's own research team
or commissioned from external consultants. It typically examines questions which are long term in nature and rarely relates to individual development projects.

This type of research is often employed where a development or invest-ment company is either reviewing its current strategy or thinking of moving in new directions. The question might, for instance, be whether the company should be moving into a different property type or location and/or whether it should diversify its activities. Reports of this nature typically compare performance across sectors and/or places over the long term; in property, this usually means the mid-1970s when property data began to be available. The analysis will usually place property market performance within the context of underlying economic forces and will try to step back from the concerns of the moment. Analysts may also draw on economic and property market forecasts to predict future per-formance. This type of strategic research is rarely published and we have little idea of how widespread its use is and its impact on the behaviour of major property companies.

Strategic research is also used to keep companies informed of emerging and future trends in occupier, development and investment markets. Research of this nature is sometimes funded by a group of developers and/or investors joining together to commission analysis on a particular topic. While little of this research is published, we know that large prop-erty companies and investors are considering issues such as 'the grey wave' as it affects retailing and the potential impact of changes in work-ing practices on the demand for office space.

Some developers use this type of research to spot new market niches. One strong proponent of such research during the 1980s development boom was Stanhope Properties. The design of its Broadgate Development in the City was heavily influenced by extensive research on occupier trends which the company had commissioned. Amongst other things, the research identified the potential impact of financial deregula-tion on City-based occupiers and the subsequent explosion of demand for buildings with large floor plates. More recently, Stanhope's research has identified a niche for what it calls 'Commerce Parks' – high quality, mixed-use business parks with an emphasis on value for money, provid-ing an alternative to the sometimes extravagant designs of the office-dominated business parks of the 1980s (Stanhope Properties, 1994).

(b) Site-specific analysis

Site-specific analysis is employed where a developer or investor needs answers to questions about an individual scheme. These questions might include, for example, whether the developer's hunches about space short-ages are correct, whether to refurbish an existing scheme or whether an investor should fund a particular scheme. The purpose of the site-specific study is to look beyond the immediate state of the market (on which agents can advise) and to identify risks and opportunities which might not otherwise be apparent. Among other things, the analyst should be able to place perceived market shortages within the context of the property cycle and to identify how the market might move during the course of the development project. Above all, property market research should be dispassionate. An especially important role for research is in providing the developer with checks and balances at times of market euphoria and, equally, in being able to identify the up-side of markets which are presently in the doldrums.

Site-specific studies typically examine some or all of the following:

i.   Demand – quantity and type
ii. Supply – quantity and type
iii. Market conditions

(i) **Demand**   The first step in any site-specific study is to define the scheme's most likely market area. This is often referred to as 'the catchment area', especially in research relating to retail schemes. The size of the catchment area will normally depend on the size of the scheme or the population of the town on which it is dependent and the expected area of 'draw' in terms of demand. Generally, the larger the development, the more extensive is the catchment area. Where a scheme is designed for occupiers who are likely to be long distance relocations, e.g. a large business park, the analyst may research demand on a country-wide or even international scale, in addition to the local market area. Demand is analysed both in terms of underlying economic drivers and recent property market transactions within the catchment area.

An understanding of the local economy is important in identifying any potential long-term threats or opportunities, as well as the potential size of the market. Very simply, since employment has a direct influence on consumers' spending potential and it is directly related to how much floorspace is needed in offices, factories and warehouses, an appraisal of the current and future jobs situation is crucial in any study. An analysis of economic structure and recent performance might, for example, reveal that the economy is over-dependent on a particular industry whose outlook is poor. Change in the size and composition of the local population will also influence spending potential and the available workforce. This local economy research will draw on the various sources of 'supporting information' outlined in Section 7.4.1 and will normally also include visits to the area and contact with organizations such as the local TEC, the local authority and, perhaps, key employers. Research of this type has been used extensively in the conception of Blue Circle Properties' 'Eurocity' development in North West Kent, a good example of a research-led development scheme (see Case Study at the end of this chapter).

A number of methodologies have been developed to translate actual and forecast volumes of spending, or numbers of jobs, in an area into the quantity of floorspace needed – or the capacity of the market. This capacity can be compared against existing floorspace provision, together with any additional development proposals, to assess whether there is likely to be a quantitative shortfall in supply, now or in the future. Establishing the quantity of demand is, however, rarely sufficient to provide a case for or against a particular development. Once the total potential pool of demand has been established, some assessment needs to be made of its qualitative characteristics and how far these match the developer's proposed design. Clearly, not all shoppers will buy identical goods or in the same stores; not all office occupiers will want the same configuration and specification of space. The purpose of looking at the 'shape' of demand can be twofold: firstly, to estimate what proportion of the total quantity of demand is likely to be for this particular type of scheme; secondly, to assess critically the design of a proposed (or existing) scheme. Qualitative aspects of demand can be assessed either via further desk research or via direct surveys of occupiers and, in the case of retail, shoppers.

Examples of desk research include: the socio-demographic composition of the population; the number of banking, legal or accountancy offices; or the size distribution of manufacturing firms. The methods are many and will depend on the research question(s) being asked. Data on property transactions can also be useful indicators of the 'shape' of demand, e.g. the size, location and rents achieved in recent office lettings, or the profile of retailer enquiries. While recent lettings can be a useful indicator of demand, a crucial distinction must be made between effective demand (completed transactions) and latent demand (requirements yet to be satisfied). In certain circumstances – in buoyant markets especially – the former may be unduly influenced by what developers are currently providing rather than what occupiers will actually want in the long term.

Surveys are typically conducted via direct interview, either face-to-face, by telephone or by post, depending on the complexity of the questions being asked. Surveys are carried out either by property researchers or subcontracted to specialist market research firms, especially in the case of shopper surveys for which the required sample size is normally large.

Surveys reduce the need to infer occupier behaviour from sometimes weak secondary data sources, such as reported deals. They can provide excellent first-hand information on occupiers' real requirements and the reasons behind their property decisions. The principal disadvantages of surveys are that they can be very costly and poor survey design can produce misleading results.

(ii) Supply   Analysis of supply can similarly be divided into quantitative and qualitative components. Quantitative analysis will typically establish the amount of existing floorspace in the market area, any development proposals and when they are likely to be built. This analysis should be placed with the context of the development cycle to indicate whether increasing or decreasing building activity is probable in the next time period. Qualitative analysis may assess how much of the total floorspace is likely to be competing directly with the client's scheme, given its characteristics and intended occupier profile. In this way it is sometimes possible to indicate gaps in provision in apparently oversupplied markets and vice versa.

As we discussed in the section on Information, data on floorspace and development activity are not readily available, especially at the local level. Most usually, the researcher has to piece together evidence from various sources to provide a best estimate of supply. Sources include local authorities (though tighter financial controls are resulting in declining data provision), commercial databases, direct field work, local agents and the developers themselves.

(iii) Market conditions   The balance between demand and supply should, of course, be reflected in market conditions, specifically in rents. Property researchers frequently use the prevailing level of vacancies as an indicator of current market balance. Where they are available, data on trends in take-up and availability are also used to indicate the direction in which market balance is moving and why. A falling vacancy rate, for example, may not always indicate improving demand; it may equally indicate the withdrawal of vacant stock from the market, perhaps to be redeveloped later, in conditions of weak or declining demand. Unfortunately, lack of data means that detailed analysis of take-up and availability is rarely feasible except where the analyst can devote considerable effort to piecing together lettings and vacancy histories.

The analyst, therefore, must marshall many disparate indicators to present a view of future market conditions and the likely outcome for an individual development. Where extensive data are available (e.g. in the

City of London) projections of take-up, availability and vacancies can be used alongside projections of underlying demand and supply conditions to produce a formal forecast of rents. Elsewhere, rents may be projected simply on the basis of underlying economic demand conditions and the supply of new developments (i.e. at one step removed from the market indicators of take-up and availability). Assumptions about future rents contained in the formal development appraisal can then be tested against the analyst's rental projections, further helping the developer to come to a view about the viability of the development project.

Researchers can also provide a more qualitative view of the viability of a scheme, pointing to intrinsic strengths and weaknesses, any unforeseen opportunities and any future threats (sometimes called SWOT analysis). This type of analysis is especially useful in situations in which formal forecasting is not possible and in addressing questions such as the optimum size and configuration of development. Conclusions from qualitative analysis of this type can also begin to identify the characteristics of the target market; occupier profiles thus identified can be used to select companies for inclusion on a marketing contacts database. The promotion of development schemes is covered more fully in Chapter 8.

7.4.3   Forecasting

As they have become more familiar with property research, developers and investors have become increasingly interested in the possibility of predicting future market conditions in a formal way – usually rents, yields, capital values and returns. To meet the demand, a number of consultancies now produce regular forecasts of the property market, usually on a confidential basis to 'clubs' of developers and investors. Forecasts are rarely published in any detail. Forecasts are typically available for the office, retail and industrial sectors as well as the investment market at national, regional and, increasingly, town level. A number of institutions' research teams also produce property market forecasts for use by their own surveyors.

Forecasts are used for two principal purposes. Firstly, they can provide an indication of the likely future operating environment through forecasts of the property market as a whole. In this respect, forecasts can indicate a required change in direction for a company's strategy, e.g. by highlighting future risks in currently buoyant markets or opportunities in presently slack markets. Secondly, as we have seen above, forecasts are sometimes employed in site-specific studies to provide rental projections (and, more rarely, yield forecasts) for the local market into which a scheme will be pitched.

Forecasts seek to predict changes in rents, yields, etc., on the basis that the relationships between each of these variables and the factors which drive them – demand, supply, market balance, etc. – will remain the same in the future as they have been in the past. Most forecasters employ formal econometric modelling techniques based on various forms of regression analysis.

As we have seen in our above discussion on Information, direct measurement of demand and supply in the property market remains extremely difficult. Forecasting models, therefore, normally employ proxies for demand and supply. On the demand-side, the economic factors which drive property demand are often used as the demand variable, e.g. retail sales or employment in service industries. On the supply-side, floorspace stock and building completions are sometimes known (e.g. in the City office market) but often they have to be inferred from data such as New Construction Orders.

While the form of individual consultants' forecasting models are closely guarded secrets, some indication of how property market relationships can be quantified in this way is available in the recent Royal Institution of Chartered Surveyors report *Understanding the Property Cycle* (Royal Institution of Chartered Surveyors, 1994a). Taking as an example the analysis of national retail rents (Figure 7.2), the research found that recent rental levels determine around 60% of the current market level, the remainder being explained by market conditions. This, say the researchers:

> reflect[s] the common sense observation that rental values do not often move wildly from high to low levels, but adjust in a series of annual steps.

Hence, current rents are strongly related to rents in the recent past. In addition to recent rents, however, a number of other factors 'explain' retail market rents, including consumer expenditure on the demand side, together with floorspace stock and construction starts on the supply side, and interest rates. A common feature of such models is that some explanatory variables are 'lagged', that is to say that events one or a number of years ago may be reflected in rents this year.

These types of explanatory model are the basic building blocks of property market forecasts. Once the fundamental relationships have been established, forecasts of the independent variables – rents, yields, etc. – can be produced if projections are available for the explanatory variables.

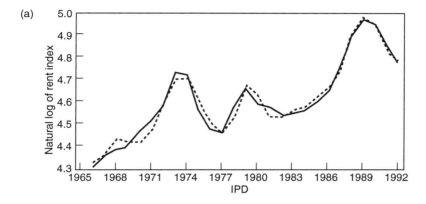

Figure 7.2   Econometric modelling of national retail rent. (a) Regression analysis: fitted (solid line) and actual (dashed line). (b) Regression equation. Reproduced with permission from *Understanding the Property Cycle*, the Royal Institution of Chartered Surveyors.

On the demand side, economic forecasts are typically obtained from specialist economic and regional forecasters, such as Henley Centre, Cambridge Econometrics and Business Strategies Ltd. On the supply side, researchers may provide their own estimates of future building and stock from analysis of the current development pipeline. In some circumstances, it is also possible to use data on take-up and availability in forecasting models.

As the use of property market forecasting has become more widespread so have the dangers of overly mechanistic analysis grown. A number of simple rules can be applied to the process of forecasting, however, to avoid some of the potential pitfalls. First, the old axiom 'garbage in–garbage out' should be applied rigorously to property market forecasts. This remains an issue in spite of the vast improvements in the quality of property data which have taken place in recent years. Second, and reflecting the restricted availability of suitable data, models should be kept simple and robust. Third, 'answers' generated by forecasts should be assessed critically in the light of the experience and knowledge of the researcher and the client, as would be any other evidence. Fourth, clients must be made aware of the uncertainties surrounding forecasts and the limits imposed by the data. At the end of the day, forecasting is not a substitute for interpretation; it is merely a further tool for the analyst to work with.

7.4.4   Portfolio analysis

A further application of property research techniques is in the area of portfolio analysis. Portfolio analysis was originally developed for the other capital markets in the 1950s but has only recently been applied to property. It is used to analyse the comparative return and risk performance of:

1. Property as a whole against other investment media (equities and gilts).
2. An individual property type against the others (offices, retail, industrials).

Very simply, the expected relative returns for different property types (or asset classes) and their volatility are identified, based on long time-series data. The analysis thereby identifies which mix of property types (or asset class) can provide the maximum return for any given level of risk, and therefore the optimal mix within a portfolio. Forecasts are increasingly being used alongside portfolio analysis to predict the optimum property mix in given future market scenarios.

A number of problems have been identified in translating portfolio theory from other markets directly to property, e.g. the special nature of the property 'product' and the poverty of property market data. Further discussion of the techniques and problems can be found in the work by MacGregor *et al.* at Aberdeen University and in other specialist texts (e.g. Brown, 1991).

This is only a brief introduction to what is a rapidly developing specialism in property research. The purpose of including portfolio analysis in this chapter has been to illustrate the full scope of what property market research is and does. Readers wanting more information on portfolio analysis are referred to the specialist literature mentioned above.

7.5   THE IMPACT OF RESEARCH

So far, we have been concerned with the description of what property research is, who does it and how it is done. We have seen that it has grown very rapidly and become an important part of the development and investment process. But how useful is property research?

Clearly, developers and investors find research helpful at the level of individual projects, if the rapid expansion of such research can be taken as a measure of its usefulness. At the level of the property market as a whole, however, the answer is less clear-cut. In economic theory, wider availability of information should help to make the market more efficient. This was evidently not the case in the late 1980s boom and bust cycle, which occurred against a background of massively increased access to property data. Indeed, Richard Barras of Property Market Analysis has argued (Barras, 1994):

> Paradoxically it [the potentially destabilizing effect of institutional investment] may be made more so by the growth of property information and research over the past decade, which has improved the market signals and therefore sharpened, not dampened, the speculative response.

However, as he and others have argued, the property market is inherently unstable by virtue of the special nature of the product (large, indivisible, locationally specific) and its production (development) cycle. Better information has merely made responses more acute. This may not be a problem in markets in which the supply side can respond quickly to changes in demand – either by increasing or decreasing output – but in the property market the consequences can be disastrous because of the long lead times in the development cycle. Against this background, we should recognize that there are limits to the usefulness of research; it can reduce uncertainty but not eliminate it altogether. As Barras points out in the same paper:

> The current condition of the market in mid-1994 perfectly illustrates the conundrum facing investors and developers, and points to why it is facile simply to ask them to 'learn from past mistakes'.

The most difficult problem they face is not to anticipate future demand (which is where research can help), but to anticipate the reaction of their competitors to the same demand signals (which is where research cannot help).

Nonetheless, as we look forward to another development upswing in a relatively uncertain economic climate the use of research will undoubtedly continue to expand. As the property research industry matures we can anticipate further improvements in the quality and consistency of data, perhaps also a better balance between information gathering and analysis, along the lines suggested in the *Mallinson Report* (Royal Institution of Chartered Surveyors, 1994b): 'Databases must be built to which all contribute and all have access. This will call for recognition that in a modern knowledge-based industry you are not differentiated by what you know, but by how you use it.'

Better databases and more consistent analysis of them, however, will not provide all the answers to the property market. We should never lose sight of the all important final consumers of property – the occupiers. Perhaps more research in future should be directed to understanding exactly what it is they require from the development industry if we are to avoid a repeat of the mistakes of the 1980s.

Executive Summary

The fast growing area of property market research embraces a number of different specialisms including information and database services, strategic and site-specific analysis, forecasting, and portfolio analysis.

While significant advances have been made in the availability of property market information, a number of major gaps persist. The application of GISs to the property market promises more sophisticated and widely available data in future.

Site-specific and strategic analysis are used by developers and investors to help assess the viability of individual projects and/or to inform the company's long-term property strategy.

Analysts provide insight on the fundamental demand and supply factors underlying market conditions; research can also pinpoint risks and opportunities which might not be readily evident from current transactions in the market.

Forecasts and portfolio analysis are increasingly being used in these types of analysis, in particular to provide predictions of future rents and yields and the optimal portfolio mix of properties as a result.

REFERENCES

Barras, R. (1994) Property and the economic cycle: building cycles revisited. *Journal of Property Research*, 11, 183–197.

Brown, G. (1991) *Property Investment and the Capital Markets*, E & FN Spon, London.

Royal Institution of Chartered Surveyors (1994a) *Understanding the Property Cycle*, Royal Institution of Chartered Surveyors, London.

Royal Institution of Chartered Surveyors (1994b) *The Mallinson Report. Report of the President's Working Party on Commercial Property Valuations*, Royal Institution of Chartered Surveyors, London.

Stanhope Properties (1994) *Commerce Parks: Meeting the Needs of Modern Businesses*, A Stanhope Position Paper, Unpublished.

CHAPTER 7 CASE STUDY – EUROCITY, EBBSFLEET & BLUE WATER PARK, NORTH WEST KENT (BLUE CIRCLE PROPERTIES)

Blue Circle has been involved in North West Kent for many years through its cement quarrying and related industrial activities. The company is now promoting major mixed-use developments in the area which will redevelop worked-out quarries and associated land. Blue Circle have already started to build the Crossways Business Park, strategically located close to the Dartford Bridge on the M25, with further development planned. Also planned is the 1.65 million sq ft Blue Water regional shopping centre, due to commence in 1995–96. In addition, the proposed development of a 'Eurocity' (incorporating Blue Water and the new international passenger rail station at Ebbsfleet) will be one of the landmark schemes within the 'Thames Gateway' – the area of north Kent and east London identified by the government as a priority for economic development. Extensive new housing and community services will complete what is envisaged as a major new settlement on the south bank of the Thames. Extensive research has helped Blue Circle to identify these alternative uses for its land as well as demonstrating the benefits of locating an international passenger rail station in the area. Research has focused on the optimum size and configuration of development given the expected growth and characteristics of the local and subregional economies. The multi-disciplinary research effort has involved the following types of consultant:

1. Planning and urban design consultants who, amongst other things, have identified the suitability of sites in the area for different uses and the number of new jobs which could be accommodated as a result.
2. Environmental consultants who have indicated the potential environmental impact of different development mixes.
3. Transport and engineering consultants who have researched the likely traffic impact of such major developments and indicated required road and public transport improvements.
4. Economic consultants who have explored future economic scenarios for the area and identified the scale of growth likely in the next 30 years, including the number of new jobs and in what industries.
5. Property market consultants who have translated the forecasts of new jobs into projections of demand for new space of different types. Property market research has been used in the assessment of the viability of the Blue Water regional shopping centre, and to indicate the likely characteristics of its shoppers and requirements of retailers.

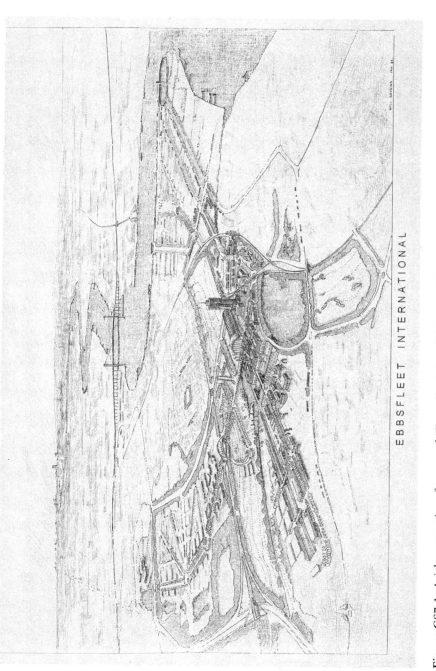

EBBSFLEET INTERNATIONAL

Figure CS7.1 Ariel perspective of proposed "Euro City", North Kent – Ebbsfleet International.

8

PROMOTION AND SELLING

8.1 INTRODUCTION

Promotion of a development scheme in order to achieve its disposal is an essential and integral part of the development process. The disposal of a building, whether through a sale or a letting, should be thought about during the evaluation process and not just a few weeks before the completion of the development. However, despite its importance, the promotional budget for a development scheme tends to be the first cost which is looked at when attempts are made to either improve the profitability of the scheme or the residual value of a scheme during the evaluation process. Also, the term 'marketing' is very often mistakenly used instead of 'promotion' to describe the process of informing and attracting potential tenants or buyers. Promotion should be aimed at the market identified by the market research, which the developer should carry out as part of the evaluation process (see Chapter 7). Without a clear understanding of the potential market and its needs, any promotion of a development may turn out to be a waste of time and money.

A developer will usually employ an agent or agents, unless an in-house sales team is employed, to secure sales and/or lettings of the scheme. Whoever is employed should be involved in the promotional and lettings/sales strategy from the beginning. Unless a scheme has been pre-let or pre-sold, an early decision needs to be made by the developer as to the method of disposal, and this will largely depend on the motivation of the developer and the prevailing investment market. Does the developer intend to retain the completed and let development as an investment or will it be sold to an investor to realize capital? Depending on the nature of the development, will the freehold or the long leasehold, of either the entire scheme or individual self-contained units, be available to owner-occupiers?

Once potential tenants or purchasers have expressed interest in response to a promotion, then the developer, in conjunction with any agent appointed, must persuade by negotiation the interested party to purchase or lease the property at the right price and on the right terms. The disposal of a development is, therefore, a two-stage process: promotion and selling.

Unless a pre-let or pre-sale has been agreed then the disposal, in terms of its timing and the rent/price achieved, will ultimately determine the financial outcome of the scheme from the developer's point of view. If the developer intends to dispose of the development on completion of the letting as an investment property then the process has three stages: the building has to be 'sold' twice, once to the tenant and then to the investor. In these circumstances, the developer has to ensure that the quality of the tenant and the terms of the letting are acceptable to the investment market (see Chapter 4).

In this chapter, we will discuss the various ways of promoting and selling a particular development, together with the role of the agent at every stage. We will conclude with a brief explanation as to how a sale/letting is completed.

8.2 METHODS OF PROMOTION

The aim of any method used to promote a development is to communicate with the potential purchasers and tenants. The developer should establish at the outset a realistic budget to promote a building and be prepared to increase it should it prove necessary. The size of the budget will depend on the nature of the scheme and the state of the occupier market prevailing at the time. There are no hard and fast rules as to how much of the development budget should be allocated to promotion: it is a judgement for the developer to make based on past experience. Developers, who specialize in a particular type of development (a good example is out-of-town retailing) or who concentrate their development activities locally, will tend to have established personal contacts and a good reputation with potential occupiers, so the amount of promotion required will be minimal. Such developers will promote the scheme by personal contact either directly or through appointed agents. However, in the case of a large town centre scheme or a large office development, promotional campaigns tend to be very extensive and may have to extend over considerable periods of time.

Whatever the size of the promotional budget, it is important that the various methods used within any campaign are monitored throughout. Research into the source of enquiries will help to gauge the effectiveness of the impact of the promotion on potential occupiers. It is important to obtain maximum value for money and a careful analysis of all enquiries in terms of both numbers and 'quality' will help to identify the most effective methods of promotion. Careful recording and analysis of results will provide a useful database for the planning and direction of future promotional activities.

The planning of any promotional campaign must be carried out in consultation with the team who will be involved in selling/letting the scheme, whether they be appointed agents or an in-house team (see Section 8.3). In addition, the team may include an advertising agency and a public relations consultant who are experienced in the promotion of property. Before any promotional activity takes place, the developer, agent and any consultants must identify the strengths and weaknesses of the particular property, in the context of the competition, so the advantages may be highlighted. They should establish during the evaluation and design process, through their own knowledge and experience, and where necessary market research, the target market for the property. The property market is very fragmented: occupiers range from small businesses to major companies with their own retained agents or internal property departments, and they have different needs and perceptions. Therefore, any promotion must be specifically targeted to be cost-effective.

One of the first decisions is the naming of the particular building or development: the name of the building, together with an associated design/logo, will last throughout the promotional campaign. It identifies the building or development, which is particularly important in relation to a large shopping centre development as the name will endure beyond the letting of the individual shop units, and it will be used in promoting the centre to the shopping public. With an office building the name will tend to change when let to a large individual firm who may name it after their company name. Names of buildings may reflect their former historical significance or their particular location. It is important that the developer, before finally deciding on a name, checks with the local authority and the Post Office that the name is acceptable to them, so that it may be used in the postal address.

The various means of promoting a property include the following.

8.2.1   Site boards and site hoardings

One of the first methods of promotion, which can be arranged as soon as the developer has acquired the site, is the erection of a site board. Site boards can be very effective and provide an inexpensive way of communicating the existence of a scheme even before there is any physical sign of it.

Several site boards may be erected during the course of the development depending on the information available at any particular time. For example, before the final details of the scheme are known, a simple 'all enquiries' board may be erected providing the name of the developer or an agent and their telephone number. This may encourage early interest but must be replaced at the earliest opportunity by a board providing information on the nature and size of the scheme. Most development sites will have a board giving details of the development, both during the course of construction and from the time the building has been completed until sales or lettings have taken place.

Great care should be given to the positioning of the board and the way it is worded to give maximum advertisement value. If the site is in a prominent location it may be the main method of promotion and may generate the most enquiries. The board should be clean, lit (if possible so as to be seen at night and in dull weather conditions) and regularly maintained. A shoddy board casts doubt upon the quality of the development itself. Care should be taken to position the board away from trees, which may obscure the board when they are in full leaf. Also the developer should ensure that the boards of contractors and the professional team are kept to a minimum and do not distract from the site board. If the developer sells advertising space on the site hoardings to advertising agencies then care must be taken to ensure that their positioning does not distract from the main site board.

It must be remembered that the site board is an advertisement and therefore requires planning permission prior to its erection. Planning is usually forthcoming fairly quickly as it is a matter which can be dealt with under a planning officer's delegated powers (see Chapter 5). Some planning authorities may impose size restrictions, e.g. the City of Westminster in London currently operates a policy which not only restricts size but also restricts the colours used to black and white.

The message on the board should be simple and uncluttered with irrelevant information. The board should inform the passer-by at a glance of the following information:

1. The name of the scheme (incorporating any logo).
2. The type and size (or range of sizes) of the accommodation available.
3. Whether the accommodation is 'to let' or 'for sale'.

4. The name and telephone number of the contact who can provide further information, either the developer or the appointed agents.

Further information, although not always essential, may include a completion date and brief specification points highlighting the existence of, for example, air conditioning and car parking. If further information needs to be added, e.g. the announcement of the letting or sale of some of the accommodation, a coloured flash containing the message can be placed diagonally across one of the top corners of the board.

A trend developed in the 1980s is the use of decorative site hoardings to surround the development site during construction to promote the scheme. Decorative site hoardings have two advantages: firstly, they distract from the unsightly chaos of building sites, providing an attractive outlook to passers by, and, secondly, they can provide information about the scheme in an imaginative and humorous way to help the awareness of the development. People will remember the scheme with the decorative hoardings rather than the usual contractor's painted plywood effort. However, they can be costly when compared with ordinary site hoardings and they also lose the benefit to the developer of revenue from advertising hoardings. In the case of a refurbishment, where protection of the building during construction is usually in the form of mesh hung from the scaffolding, it is possible to reproduce a life size image (or even a promotional message) of the proposed building onto the protective mesh.

8.2.2   Particulars and brochures

Any promotion will need to be supported by particulars and/or brochures of the scheme providing the potential occupier with further information in response to their enquiries. Particulars need to be aimed at potential occupiers and agents. The nature of the particulars will depend upon the nature of the property and the target market identified by market research. The fundamental question that needs to be asked is what information will the reader require in order to encourage them to arrange to view the property. The content and nature of the particulars and brochures, therefore, need to be aimed at the likely readership. For example, on a small industrial scheme, the target market may be small businesses. On a major town shopping scheme, the target market may be a combination of local retailers and the major multiples with their retained agents.

Particulars in some cases, usually on smaller schemes, are set out on a single sheet of paper with a photograph or on a double sided printed colour glossy sheet. In other cases, on larger schemes, more detailed information is needed and particulars need to take the form of glossy colour brochures or booklets.

Simple particulars, which may be supported by photographs or illustrations, should contain the following:

1. A description of the location of the property.
2. A description of the accommodation, giving dimensions and a brief specification.
3. A description of the services supplied to the property such as gas and electricity.
4. The nature of the interest that is being sold if it is freehold, the nature of any leases or restrictions to which it is subject, and, if leasehold, the length and terms of the lease.

The particulars should give the name, address and telephone number of the agent and developer, and the name of the person in the agent's office or on the developer's staff who is dealing with the property, so that a direct contact can be made by the potential occupier.

Care must be taken to ensure that all information given in the particulars is accurate and not 'false or misleading', in order to comply with the Property Misdescriptions Act 1991. Therefore, all information needs to be carefully checked to ensure that the particulars are accurate as far as can be reasonably ascertained. This rules out the use of 'flowery' or subjective language in the description of the property. The Act applies to any information supplied or statements made by agents and developers in promoting the property: so it applies to all forms of promotion.

In most cases, it is necessary to prepare a colour brochure describing the property, giving more details than the brief particulars referred to above. The brochure is sent to the people who reply to any advertising, site boards, mail shots or preliminary particulars, or may be sent directly to people who are known to be interested. Therefore, it is important that the brochure is integrated with the rest of the promotional activities. The appearance is of great importance as it reflects the quality of the development that it describes. Unless the developers or their agents have in-house specialist staff, it is best to employ an outside designer or agency with experience in property brochures. They can provide professional designers and copywriters, ensuring the most effective use of words, typography, colour, shape, etc. First impressions count and the brochure is often the first introduction to a building for a prospective occupier. Therefore, it is important to ensure high standards of design and production.

The brochure should include maps, plans, photographs and a brief specification. All photographs need to be taken professionally and many photographers specialize in property. The brochure should provide clear directions and maps to enable the reader to find the property easily. It also needs to provide information about the locality which will be helpful to a potential occupier. For example, in promoting a particular office location information should be provided on drive-times to nearby motor-ways/roads, towns, airports and cities; train services to London and major cities/towns nearby; and the availability of public transport services. Also, particularly if the office is in a provincial town and the likely occupier is a firm decentralizing from central London, information should be provided on local housing, shopping facilities, availability of skilled labour and so on.

In a lengthy promotional campaign it may be necessary to produce more than one brochure at different stages in the development. Often much time and money is spent on brochures; in addition to being expensive to produce, they are sometimes quickly out of date. At first, simple particulars or what is known as a 'flyer brochure' (a single sheet brochure) may be produced showing an artist's perspective of the proposed building. These will need to be replaced as soon as possible with photographs of the completed building as they are available. Where changes in the development scheme take place and the promotional budget permits, those elements that are likely to change, such as floor areas and perspectives, photographs can be incorporated in loose-leaf form, or, better still, the brochure can be bound with binders, enabling pages to be updated at ease without the expense of producing a new brochure.

Also it may be worthwhile producing a tenants guide to the building for prospective occupiers who show continued interest in the building. The guide should provide a detailed description of the building and specification, together with information on maintenance and energy-saving data.

On particularly large and complex development schemes a video (or the latest advance in computer technology the 'CD ROM') is often used as a promotional tool in conjunction with the brochure. With a CD ROM graphics can be combined with video and voice in a multi-media presentation. However, videos and CD ROMs are expensive to produce and can become quickly out of date if used too early on in the development process. Videos are particularly useful in show suites or offices where they can be shown to interested parties who visit the site. Most importantly, they can most effectively bring the scheme to a potential occupier without them having to leave their office. A slide video presentation is an alternative; it has the advantage of being less expensive and the material can be easily updated.

In the early stages of the development, prospective occupiers may be shown three-dimensional images of the building proposed if the architect is using a CAD (computer aided design) system to produce drawings of the scheme. This technology enables the viewer of the computer screen to be effectively taken round the building from every angle and perspective. Photographic stills of the three-dimensional images produced can be also be used in any promotional material.

8.2.3   Advertisements

Advertisements are either aimed directly at potential occupiers or indirectly through their agents depending on the target market Such advertisements are usually placed in the property sections of local, and where appropriate national newspapers; in the various property journals, such as *Estates Gazette*, *Property Weekly* and *Estates Times*; or, where appropriate, in special trade journals. If the target market is likely to be local businesses then local newspapers are the most obvious place for advertisements. If the potential occupier is likely to be represented by an agent, then the property press is the most effective place for advertisements.

The content of an advertisement is obviously important, but unquestionably its design and layout are crucial. Good design and layout increase the impact of the advertisement, whereas poor design not only wastes money, but can sometimes create an adverse impression. Poor design should be avoided at all costs. It must be remembered, particularly in the property press, that the advert will be one of many and therefore must be designed to make an impression. The cost will vary with size and position in the paper or magazine. A front page while costly may provide the opportunity to create a bold, imaginative advert which will make a lasting impression. The cost of a one-off advertisement on the front page has to be weighed up against the cost of a series of smaller adverts in less prestigious positions. The style and advertising media should be varied in a constant attempt to improve results.

Great care needs to be taken over the way in which the advertisement is worded and presented and in nearly all cases it is worth employing an experienced advertising agency to ensure the maximum impact, although some large firms of agents have their own in-house advertising department. The advertisement should contain sufficient information to attract the potential occupier, but it should not be so cluttered that it is difficult to read. Information should include:

1. The type of property.
2. Approximate size.
3. Location.
4. Whether the property is for sale or to let.
5. The telephone number of the agent or the developer, or both, should be given, so that the potential occupier can make a direct contact.

It is always difficult to judge the effectiveness and worth of an advertising campaign, but some attempt should be made to identify the source of any enquiry. Of course, it must be remembered that some readers might see an advert and take no immediate action, but the information will stick in the memory, and subsequently looking for accommodation, they will recall the advertisement and make enquiries.

Newspaper and magazine advertising is far and away the most common form of advertising that is used. Very little use has so far been made of TV and radio for individual properties, although there has been an increase in the use of this medium as it becomes more cost effective with increased competition between media companies. Conventional wisdom seems to dictate that radio and TV is only used when all else has failed, although advertisements on local commercial radio during 'drive-time' are often effective.

Poster advertising is also a medium which is increasingly used to promote larger properties, employing prominent locations such as railway stations, roadside hoardings, and even on or within public transport and taxis. As an example of its effective use, consider the promotion of an out-of-town office development with ample car parking. Posters displayed in appropriate railway stations, in the underground, on buses or in taxis in central London can be effective means of advertising to those who are forced to travel in crowded, uncomfortable conditions into the capital. A poster advert, like a site board, should convey its message instantaneously as people will only have time to glance at it and rarely to study it carefully. However, poster advertising is expensive and is difficult to target specifically.

8.2.4   Mail shots

Direct mail shots may be effective and relatively cost-effective. Mail shots are used on their own or to support an advertising campaign. However, because it has become such a widespread promotional method great care has to be taken if it is to be really effective. They should be aimed at a very carefully selected list of potential occupiers and the success of the shot will depend very much on the compilation of the mailing list, and any follow up procedures. There are a number of specialist direct mail organizations capable of producing mailing-lists with a degree of specialization and accuracy. They maintain general lists of industrial or commercial firms which can then be broken down into particular categories such as company size, trade and location. There is a limit to the frequency with which direct mail shots can be used and the employment of one of the specialist firms will often be found to be advantageous. Care must be taken in selecting firms: they must routinely and frequently update their mailing lists. To be effective the mail shot must reach the right person within the organization, who is responsible for property matters, and at the right address. To overcome the almost automatic tendency of the recipient to throw the contents of the direct mail shot into the waste-paper basket, make sure that the message can be seen at a glance. A very long letter which has to be read through to the end before the message is fully understood, or a brochure or leaflet without any covering message, will often fail. By contrast, a short sharp covering message attached to a brochure or leaflet will often be absorbed by the recipient before they have time to throw it away. The letters should be prepared in such a way that each one appears to have an original signature.

With mail shots, unlike advertising, the results and effectiveness of any promotion can be quantified. It can be also followed up by a telephone call to obtain a reaction or information on general property requirements. Such telephone canvassing may be carried by the agents or a representative employed in the show office/suite. Mail shots, provided that the target market is accurately identified, can be cost-effective.

8.2.5 Launching ceremonies

In some cases, it may be appropriate to launch the development by a topping-out ceremony during the course of construction or by an opening ceremony when the building is completed. It is normal to invite to such functions the local and national press, local and national agents, and sometimes potential occupiers. Local councillors or their officers may also be invited to help with the development company's public relations. Types of ceremony vary enormously but normally involve some kind of refreshment, perhaps a buffet lunch, centring on a short speech by the developer or maybe the mayor or local MP. The opening of a shopping centre may be quite a grand affair with a celebrity to perform the opening ceremony. Entertainment should be imaginative with a definite theme, perhaps extending from fancy dress to the type of food. It is important to pay great attention to detail; a ceremony enjoyed by those who attend it will be remembered.

Whatever form of ceremony or party is devised, it is of the utmost importance that at the time of the function the development is in good order. At an opening ceremony the buildings must be ready for occupation with all services in working order. In the case of a shopping centre, the major units may be let and occupied, and the opening of the development will be linked to the opening of those units.

It seems that almost every office and large shopping development has some form of ceremony these days. Developers have had to make them increasingly imaginative and innovative in order to encourage agents to attend. Agents will receive many invitations to ceremonies and they have to make a choice about which to attend. Accordingly, such incentives as free gifts, a prize draw to win a holiday or tickets to a major sporting occasion may have to be offered. The main advantage of this type of promotion is that it brings agents, particularly those retained by a potential occupier, to the property.

8.2.6 Show suites and offices

A promotional campaign will be timed to have its maximum impact when the building is completed, so that photographs can be included in the material and potential occupiers can view the final product. However, developers will always aim to sell/let the building before completion to maximize the cash-flow position. During the construction period any promotional activity can only be supported by the use of plans, artist's perspectives, models and CAD images to help build up a picture of what the eventual scheme will look like. Accordingly, particularly on large, mixed-use or phased developments (e.g. a business park), it is worthwhile to set up a show office on site during the course of construction to display plans, models and promotional material of the scheme. The office, often a portacabin which has been appropriately decorated, painted and fitted out inside, should be staffed by an on-site representative of the developer or agent who is able to talk knowledgeably to agents and prospective occupiers who arrive on site. Refreshment facilities should be available to offer to the prospective purchaser/tenant. A 'finishes' board should be on display showing samples of the internal finishes proposed for the scheme. It is important to make sure that the show suite or office is clearly signposted and, if possible, the landscaping should be brought forward to enhance the environment. In any event, any landscaping in a scheme should be planted as soon as possible, allowing for planting seasons, so that it can mature slightly before completion of the whole scheme.

Once any development has been completed, its appearance is of vital importance. In the case of an office scheme or an industrial scheme with high office content, it will often be advisable to fit out a floor, or a part of a floor, as a show suite with partitions and furniture, so that potential occupiers can see clearly how the offices will look when they are occupied. This may be supported by plans showing alternative open-plan or cellular layouts. It may also be advisable to fit out the main entrance hall and reception area, and it is important to ensure that the common parts and lavatories are kept clean. In nearly all cases, carpeting will be provided by the developer throughout the building, and if lettings prove difficult, it may be that this will be extended by the developer offering to fit out or furnish an office for a particular tenant. In the case of a shopping scheme or warehouse scheme, where units are in shell form, a sales office on site is useful to welcome potential occupiers.

8.2.7   Public relations

Many developers (particularly public quoted companies) either retain PR consultants or have in-house staff responsible for relations with both the general public and the press. Anyone responsible for public relations should be involved in a particular development scheme right from the beginning and be kept informed at all times on the progress of the scheme. Although, their main priority is to promote and enhance the reputation of the development company as a corporate entity, there will be spin-offs in relation to the promotion of individual development schemes. Where a scheme is particularly sensitive in relation to local politicians and the local community it is worthwhile to employ a PR consultant specifically for that scheme.

Through their press contacts PR consultants can achieve useful press editorial at significant stages in the development programme. For example, carefully controlled and timed press releases may be made to the property press on any of the following occasions: the acquisition of the site, the obtaining of planning permission, the completion of the scheme and on the letting of all or part of the scheme. Editorial coverage is often offered by a publication if advertising space is being booked to coincide with coverage of a particular geographical area or a particular type of property. Editorial coverage is much more likely to be read than an advert and can be achieved at no cost.

Promotional material should be on display at every public relations event where either the developer or the development is being promoted, such as corporate hospitality events and exhibitions.

8.3 THE ROLE OF THE AGENT

Traditionally, the agent is responsible for selling/letting a particular property on behalf of a developer. Why do developers appoint agents instead of selling/letting the property themselves? Although some developers have an experienced in-house selling team, it is not always possible to rely solely on their efforts. The appointment of an agent will depend upon the circumstances of a particular scheme and the likely market. Agents offer certain advantages.

Agents are often in a better location for attracting business. To take an example, if a London-based developer is developing an office scheme in a provincial town, it makes sense to employ an agent with offices in the particular provincial town. A good firm of agents should have detailed knowledge of the particular market in which they operate, being thoroughly familiar with current and future levels of demand and supply. They may also have additional special knowledge and experience and may, for example, specialize in certain types of property. They will also have personal contacts with agents retained by a specific client or decision makers within the property department of a potential occupier. In order to be effective, the agent must be continuously involved in the market place, so that they are aware of changes in market conditions.

Agents can be retained on a 'sole agency' or 'joint agency' basis. As sole agent they are alone responsible for disposing of the property and are entitled to a fee on each letting or sale. A normal range of fees might be 10% of the first year's rent for a letting (allowances will be made to take account of any rent frees or inducements given to the tenant on a letting) or 2% of the sale price for a sale. A joint agency arises where two or more agents are instructed to sell the same property. This can often happen where both a national and local firm of agents are instructed together. In the example mentioned above, the developer will appoint a London agent as well if market research shows that companies relocating from London might be attracted to the scheme. In such a case, on each letting or sale, the developer will have to pay a larger fee, perhaps one and half times the normal amount, the agents sharing the fee between themselves on some agreed basis. For example, joint agents would receive 7.5% each from an agreed fee of 15% of the first year's rent for a letting instruction. The appointment of a second or third agent (if two are already appointed) is often not the decision of the developer, but a condition of a funding agreement, where the fund wishes to see their own agents involved.

It is usual to employ a national or regional firm of agents operating with or without the help of a local firm. There is no general rule to apply; each case must be looked at on its own merits, but a national agent normally has a greater understanding of the larger and more complex schemes and has more direct and frequent contacts with the larger companies and multiple retailers attracted to them. Additionally, they can offer a wider service to include investment advice if the scheme is to be forward funded or sold on completion. On the other hand, the local agent will often have a better understanding of the particular characteristics of the local market and of local occupiers and retailers.

Whichever type of agent or combination of agents is used, it is important that they can have the opportunity of contributing to the planning, design and evaluation of the project. They should be able to draw attention to features of the design that add to or detract from the marketability of the property and be able to comment on prices or levels of rent, the nature of competitive development and the most effective time for letting or selling. By being brought in at an early stage, they become thoroughly familiar with the property that they are going to sell or let. Nothing is more annoying to a potential purchaser or tenant than to find that the selling agent cannot provide full details of the property that they are offering. They should attend regular meetings with the client to be kept informed of progress on the development and to plan the promotional activities.

It is necessary to distinguish between the appointment of agents and offers simply to pay commission to an agent introducing purchasers or tenants. There is a contractual relationship with the former, the terms of which (both expressed and implied) need careful consideration. The agency agreement should make clear the length of time the appointment will remain in existence and how it may be terminated, specifying any retainer payable and setting out the rate of commission and in what circumstances and when it will be payable. In addition, it will specify whether the agreement is for a sole or joint agency and what are the developer's rights concerning their discretion to employ additional agents. It should be made clear whether agents are entitled to expenses, whether or not they succeed in disposing of the property. A little time and care spent in precisely defining the relationship can save a good deal of misunderstanding later.

On occasion, offers will be made to pay commission to agents on the signing of legal agreements for sales or lettings with purchasers or tenants introduced by them. No formal agency appointment is made in these circumstances, and this type of arrangement is significantly different from the formal appointment of agents to act for the developer. Sometimes the offer is open to anyone introducing a purchaser or tenant, sometimes the offer is made to a limited number of agents.

On the face of it, a developer might think that an offer to pay introductory commission to almost anyone would be likely to result in most business. This is not necessarily so. If too many agents are handling the property, there is a danger that it will be 'peddled around the market', thus creating an unfavourable impression on the ground that if so many agents are handling the property, it must be difficult to shift because there is something wrong with it. After an initial flush of enthusiasm, agents tend to lose interest if they know that the property is in a large number of hands, and some of the best agents might be reluctant to be involved with properties widely and indiscriminately offered in such a way. Sole or joint agents, or perhaps three or four working on an agency basis, are much more likely to have a sense of involvement. Whatever the arrangements, it is important to keep all the agents informed of them at all times.

Meticulous records must be kept, so that it can be seen at a glance which prospective purchasers or tenants have been introduced by which agents and precisely when. The question of disclosure is important. If agents are already retained to find accommodation for the prospective purchaser or tenant, they will not be able to accept an introductory commission from the developer. When agents refer to 'clients', they are normally retained and will not seek a commission; but when they refer to 'applicants', they will normally expect a commission because they are not retained by applicants. In all cases, the position should be clarified explicitly at the outset.

If agents are appointed by a developer they should be involved right from the beginning of the development process. One of the agents appointed may have even been responsible for initiating the development process by introducing the site to the developer and/or arranging the necessary development finance. They have a valuable input, through their knowledge of the both the occupier and investment market, to the evaluation and design processes. Developers should be aware that agents will always tend to be cautious when advising on design and the specification of the scheme from a letting/selling point of view as they wish to ensure that the scheme has the widest tenant/purchaser appeal. Developers should make their own judgements of the advice received, backed up by their own experience and research. More importantly both developers and agents should obtain feedback from tenants on similar schemes. Not enough of this valuable research is carried out.

Meetings with agents will normally take place on a monthly or fortnightly basis depending on the stage reached within the development programme. At these meetings the agent should report progress in relation to the letting/selling of the scheme. The report will normally take the form of a schedule detailing any enquiries received from interested parties and any known requirements for accommodation which need to be followed up. The agent should advise on the availability of competing buildings and what terms are being quoted on those buildings. All concluded sale or letting transactions on similar accommodation should be reported. The source of each enquiry should be noted so that the developer can judge the effectiveness of a particular method of promotion. Promotional activities will also be discussed and the agent should have a full input in any decision on the content and design of all promotional material. The agent will usually be responsible for the booking and scheduling of all advertisement space, unless an advertising agency has been appointed. As both agents and advertising agencies book space on a regular basis they should be able to pass on the benefits of any discounts and should be able to obtain good positions within the relevant publications. Agents will also be responsible for organizing the mail shots both to potential occupiers and other agents in the market.

8.4   SALES/LETTINGS

Once a potential occupier (referred to as an applicant) has expressed interest in viewing the property then the agent will show them round in the first instance. If continued interest is shown then the developer will tend to be involved in all future viewings and negotiations with the interested party. There is no substitute for personal contact with the applicant, although there may be times, particularly early on in the negotiations, where the agent should lead the discussion. The developer should make a judgement as to when it is appropriate to become involved.

The Spindles, Oldham – a 280 000 sq ft (26 013 sq m) town centre retail scheme developed by Burwood House Developments Ltd which opened for trading in September 1993. This photograph of the Central Square shows the two anchor tenants Debenhams and C & A.

At their regular meetings agents will be able to advise the developer when it is appropriate to quote terms and at what level of rent. Agents should know the flexibility they have in negotiating terms, bearing in mind the requirements of the financier or the investment market. In fact, the terms of any letting will be determined by both the method of funding the scheme and the intentions of the developer as to whether it will be retained as an investment or sold to an investor. If a scheme has been forward funded then the fund will need to approve the terms of any letting and the type of tenant (see Chapter 4). If the intention of the developer is to sell the completed and let scheme to an investor then the terms of any letting must be acceptable to the investment market at which it is aimed. As we have already discussed in Chapter 4, this will depend on whether the property is considered to be 'prime' in terms of both its quality and location. The developer should be advised by the agent who will be responsible for handling the investment sale. The developer must continually bear in mind that the terms of any letting will directly affect the investment value of the completed development. Even though a property may be retained by the developer initially, thought must be given to any eventual sale in structuring the terms of a lease so that it is generally acceptable to investors.

We will now concentrate on the various terms which need to be agreed when negotiating with a potential tenant. If negotiations are being held with an owner occupier then the only terms to be agreed are the price and any work to be carried out by the developer. The terms to be conceded when agreeing a letting are as follows:

8.4.1 The demise

The accommodation being let needs to be defined and this is known as the 'demise'. A decision needs to be made, in the case of an office development, as to whether lettings of single floors or part floors are acceptable to the financier and/or the developer. In many instances, to maintain the maximum investment value, agents will be instructed to seek a single letting of a building. A multi-let building will narrow the potential investment market due to the additional management and risk involved which will be reflected in any investor's required yield. There are many examples of schemes developed in the late 1980s which are still vacant where the developer has strictly adhered to a policy of accepting only single lettings. A decision as to whether to relax such a policy and accept multiple lettings might need to be made if the letting of the building proves difficult. Flexibility will depend on the financial status of the developer and/or the influence of any financier or development partner. The question needs to be asked is it better to maintain cash-flow at the expense of maximizing the capital value?

If the demise is not self contained then rights will need to be reserved for the tenant in respect of the use of common parts, such as staircases, lifts, toilets, etc. Arrangements will also need to be made in respect of separating services for metering purposes.

8.4.2   Rent

The developer will usually quote a rental value for the accommodation in consultation with the agent based on an assessment of the open market value for the property. The developer will also calculate the rent required, which when capitalized at the appropriate yield will provide the developer with a satisfactory return. It is important for the developer and the agent to be aware of all recent letting transactions and the rents being quoted on other properties. When analysing letting deals and quoting rents allowances must be made for any differences in the specification and location of the comparable properties.

The actual rent agreed between the developer and a potential tenant will be a matter of negotiation and will depend on the strength of their respective bargaining positions in the prevailing market conditions. In difficult market conditions where tenants have considerable choice the developer may consider offering inducements in order to maintain the required rental value for the property. However, such inducements should be kept within reasonable limits otherwise the developer will be trading inducements in return for a rent in excess of a market rent: the developer will be 'over-renting' the building in the view of any potential investor. Reasonable inducements, such as 6 months rent-free (to allow time for the tenant to fit out the building), or a capital contribution towards specified fitting-out works (e.g. carpeting) are considered as normal market practice by potential investors. A fitting out contribution will usually be related to fixtures and fittings which are referred to as 'landlord's fixtures and fittings' for rent review purposes. Accordingly, when the rent is reviewed in accordance with the terms of the lease account will be taken of the benefit of those landlord's fixtures and fittings in determining the rent for that particular property.

There have been many examples, in the difficult letting market since 1989, of substantial inducements (12/18 months up to 3 year rent-free periods) being offered in order to maintain a certain rental level on office schemes. This would appear to maximize the investment value for the developer when the rent review clause in the lease is drafted on the basis of an upward-only rent, ignoring the existence of inducements being offered in the open market at the time of the rent review. However, with the background of low rental growth as there is currently, potential investors will discount the value of the slice of the rental income considered to be in excess of market rent as they see no growth until after the first review. In addition, there have been four cases in the Court of Appeal recently where tenants have questioned the validity of rent review clauses which ignore rent-free periods for the purposes of establishing the market rent at review time. In the judgement on these cases it was held that the presumption when constructing rent review clauses should be 'reality'; resulting in three of the tenants winning the argument that the rent at review should be based on open market rents which have been agreed without inducements.

8.4.3   Lease term

The length of a lease and the existence of any options to determine that lease will depend on whether the property is considered as a potential investment for an institutional purchaser. As we have already discussed in Chapter 4, institutional investors prefer 25 year leases, but shorter terms, e.g. 10–15 years, are increasingly becoming more acceptable depending on the circumstances. There is currently a trend towards shorter leases as tenants have been in a better position to negotiate such terms to suit their business plans. They have also been able to negotiate break clauses at certain specified times within their leases to allow flexibility to determine their lease. Institutions may, depending on the circumstances, accept the existence of such clauses provided they become operative after 15 years of the lease term. Developers may also negotiate a landlord's break clause to coincide with a redevelopment or refurbishment opportunity.

With forward funded arrangements the institution will dictate the length of the lease in the funding agreement and any flexibility will need to be strongly argued by the developer in the light of market conditions.

On schemes where the likely tenants are small businesses or sole traders, with little or no financial track record, e.g. a specialist shopping scheme or small industrial units, then the developer will tend to be more flexible. Such schemes due to the type of tenant will not be considered 'prime' and, therefore, institutionally acceptable lease terms are not so important.

8.4.4 The tenant

The financial status of all potential tenants will need to be checked by the developer to ensure they have the ability to pay not only the rent agreed but all outgoings on the property. As we have already discussed in Chapter 4, institutional investors usually like to see that the tenant's profits exceed a sum three times the rent. In most cases the developer will need to see at least the last 3 years' accounts of the tenant to establish the tenant's financial standing. In the case of new businesses or those without sufficient track record then the developer will be advised to obtain bank references and trade references. In addition, any business plan and cash-flow projections should be examined.

If the developer or fund consider the financial covenant of a tenant is insufficient then both bank and parent company guarantees may be sought to provide some form of guarantee in the event of non-payment of rent. With private companies and sole traders, directors' personal guarantees will be sought.

On a retail scheme the developer and/or the fund may wish to influence the tenant mix within the scheme to ensure a variety of retail uses for the shopping public, who will in turn determine the success of the scheme.

8.4.5 Repairing obligations

The responsibility for repairs and maintenance during the term of the lease will depend on whether the demise is self-contained. In the case of a single building a tenant will become responsible for all repairs both internal and external. Where the demise is only part of a building then the tenant will usually be responsible for internal repairs and the landlord will be responsible for the external and common parts. The landlord's expense in repairing, servicing and maintaining the common parts will be recovered by a service charge directly levied on all tenants, usually in proportion to the area of their demise. Developers must provide prospective tenants with estimates of the service charge.

8.4.6 Other considerations

The above is by no means an exhaustive list of the matters to be considered by the developer when negotiating with tenants but it represents the most important ones.

Any letting or sales transaction will need to be legally documented in the form of a lease or a contract and transfer of title. It is important that the developer instructs a solicitor as soon as the promotional campaign starts to prepare all the necessary draft documentation, so that once negotiations are concluded with a tenant or purchaser the legal work can proceed quickly. A developer will also need to ensure all the necessary draft deeds of collateral warranty have been agreed with the professional team and contractors (with design responsibility) should either the tenant or the purchaser wish to benefit from them. A tenant is likely to require warranties on schemes where the repairing obligations could be considerable if something goes wrong and they wish to be able to pursue the relevant professional or contractor for a remedy. As we have discussed in Chapter 6, the remedies under a warranty are limited and a tenant may wish to benefit from decennial insurance to protect against latent defects. However, the developer will need to judge the likelihood of such insurance being required by either a tenant or purchaser as the option to arrange cover must be arranged before construction starts.

Both intending tenants and purchasers will typically make enquiries through their solicitors as to the existence of services and require to see copies of all planning and statutory consents. Any tenant or purchaser will also require full information on the building in the form of 'as-built' plans and maintenance manuals. The developer should therefore ensure that all the necessary supporting paperwork is in place before draft contracts and leases are issued to reduce the risk of delays in the legal process. Overall the developer should be prepared to answer all queries and supply any document related to the development.

The development process does not end the once the documentation has been completed and the keys handed over to the tenant or purchaser. The developer will have ongoing responsibilities such as ensuring the satisfactory completion of the defects identified at practical completion and assisting the tenant with any 'teething' problems. The developer will have a continuing relationship with the tenant as their landlord unless the property is sold either for occupation or as an investment. Even if there is no continuing contractual relationship, the developer should maintain contact with all their tenants or occupiers and provide an after-sale service. It will pay future dividends and help the reputation of the developer in the market. Occupiers may not speak to each other but they may talk to their agents if they have cause to complain about a particular development. There is no substitute for direct feedback from which lessons can be learnt for future developments. The development industry as a whole should research occupiers' needs to a much greater degree than currently. The occupier is the customer and they should come first.

Executive Summary

The securing of an occupier should be at the forefront of a developer's plans at the start of the development process. Having established during the evaluation stage the existence of occupier demand for the development in the proposed location, further research should be carried out to identify the target occupier market and their requirements in terms of the design and specification. However, such research is often overlooked by developers which may result in a building being difficult to dispose of, which cannot be solved by simply throwing money at promotion. The aim of promotion is to make potential occupiers aware of the development scheme and for a campaign to be effective it must be carefully targeted at and tailored to its audience. Agents, with their knowledge of the market, have an important role to play in disposing of the development to an occupier whether through a letting or sale, together with any subsequent investment sale. Any letting should be secured on terms acceptable to the investor market to maximize its value.

APPENDIX A: TOWN AND COUNTRY PLANNING (USE CLASSES) ORDER 1987

The use of land is defined as follows:

Class A
Class A1 Shops (retail sale of goods except hot food including hairdressers, funeral directors, hire shops and dry cleaners)
Class A2 Financial and Professional Services (where professional and financial services are provided to members of the visiting public)
Class A3 Food and Drink (sale of food and drink for consumption on premises and hot food 'take aways')

Class B
Class B1 Business use for any of the following:
(a) Offices (except use within A2)
(b) Research and development
(c) Industrial process 'which can be carried out in any residential area without detriment to the amenity of that area by reason of noise, vibration, smell, fumes, smoke, soot, ash, dust or grit'
Class B2 General Industrial
Class B3 Special Industrial Group A
Class B4 Special Industrial Group B
Class B5 Special Industrial Group C
Class B6 Special Industrial Group D
Class B7 Special Industrial Group E
Class B8 Storage or Distribution

Class C
Class C1 Hotels and Hostels

Class C2 Residential Institutions (e.g. hospitals and nursing homes)
Class C3 Dwelling House

Class D
Class D1 Non-residential Institutions (e.g. schools and libraries)
Class D2 Assembly and Leisure (e.g. cinemas and bingo halls)

APPENDIX B:
PLANNING POLICY
GUIDANCE NOTES

PPG Note No.	Title	Year Issued
1	General Policies and Principles	1992
2	Green Belts	1995
3	Land for Housing	1992
4	Industrial and Commercial Development and Small Firms	1992
5	Simplified Planning Zones	1992
6	Town Centre and Retail Developments	1993
7	Countryside and the Rural Economy	1992
8	Telecommunications	1992
9	Nature Conservation	1994
12	Development Plans and Regional Planning Guidance	1992
13	Transport	1994
14	Development of Unstable Land	1992
15	Planning and the Historic Environment	1994
16	Archaeology and Planning	1992
17	Sport and Recreation	1992
18	Enforcing Planning Control	1992
19	Outdoor Advertisement Control	1992
20	Coastal Planning	1992
21	Tourism	1992
22	Renewable Sources of Energy	1993
23	Planning and Pollution Control	1994
24	Planning and Noise	1994

BIBLIOGRAPHY

This book has provided an overview of the development process at an introductory level. The following list is intended to provide interested readers wishing to study a particular aspect of the development process in greater detail with suggested further reading.

CHAPTER 1: INTRODUCTION: THE DEVELOPMENT PROCESS AND ITS ECONOMIC CONTEXT

There are a variety of views on, and descriptions of the development process:

Barrett, S. *et al.* (1978) *The Land Market and Development Process*, School for Advanced Urban Studies, University of Bristol – includes a model of the development process showing all the various internal and external pressures and influences. In particular it looks at the role of the local authorities and government agencies within the context of the process.

Goodchild, B. and Munton, R. (1985) *Development and the Landowner*, Allen & Unwin, London – includes a chapter on the relationships between the actors involved in the development process.

Moor, N. (1983) *The Planner and the Market*, George Goodwin, London – looks at the development process from the planner's viewpoint. It examines each of the different commercial and residential markets.

Punter, J. (1986) Aesthetic control within the development process. *Land Development Studies*, **3**, 197–212 – this article examines the practice of aesthetic control during the planning stage within the context of the development process based on a study of speculative office development in Reading. There is a useful diagram of the stages and actors involved, concentrating on the planning stage in detail.

Ratcliffe, J. (1978) *An Introduction to Land Administration*, Estates Gazette, London – although now out of date, an established text providing a useful description of the development process, with diagrams, in Part 3. There is also a chapter on each type of commercial development.

Studies on landownership are included in:

Massey, D. and Catalano, A. (1980) *Capital and Land*, Edward Arnold, London; and Goodchild and Munton (as above).

In relation to the economic context within which the development process operates there are two economic textbooks designed for students on courses relating to the built environment and construction:

Manser, J.E. (1994) *Economics: A Foundation Course for the Built Environment,* E & FN Spon, London.
Myers, D. (1994) *Economics and Property,* Estates Gazette, London.

At a more specific level there are several sources of reading in relation to building and economic cycles (see Chapter 7 references).

CHAPTER 2: LAND FOR DEVELOPMENT

Barrett, S. and Whitting, G (1983) *Local Authorities and Land Supply*, Occasional Paper 10, School for Advanced Urban Studies, University of Bristol – report commissioned by the Department of the Environment into the role of local authorities into the supply of development land to the private sector for development.
Goodchild, B. and Munton, R. (1985) *Development and the Landowner*, Allen & Unwin, London – examines the role of the landowner in the development process.

Readers wishing to cover the legal aspect of the land acquisition process should refer to some of the many textbooks covering land law and environmental law:

Card, R., Murdoch, J. and Schofield, P. (1990) *Law for Estate Management Students*, 3rd edn, Butterworths, London – established textbook for students with section on land law in England and Wales.
McAllister, G. (1992) *Scottish Property Law: An Introduction*, Butterworths, London – student textbook covering the law relating to landownership and conveyancing in Scotland.
Ball, S. and Bell, S. (1991) *Environmental Law*, Blackstone Press, London – for those readers requiring further detailed information on environmental law as it affects landowners.
Wilbourn, P. *et al.* (1993), *Contaminated Land*, Owlion Audio Programme (RICS and College of Estate Management) – provides a useful explanation of the Environmental Protection Act 1990.

For more information on the law and procedures of purchase:

Denyer-Green, B. (1994) *Compulsory Purchase and Compensation*, 4th edn, Estates Gazette, London.

Bowerman, F.G., Goodchild, R. and Chase, M. (1989) *Compulsory Purchase and Compensation*, Owlion Audio Programme (RICS and College of Estate Management) – explains the statutory background and rules of compensation.

For readers with an interest in urban regeneration the following provide more detailed study and analysis:

McCreal, W.S., Berry, J.N. and Deddis, W.G. (1993) *Urban Regeneration,* E & FN Spon, London – provides a very detailed analysis of the role of property development and investment in urban regeneration.
Healey, P. *et al.* (1992) *Rebuilding the City*, E & FN Spon, London – looks at property-led urban regeneration in the 1980s, examining the inter-relation between property development and urban regeneration policy.
Smyth, H. (1994) *Marketing the City*, E & FN Spon, London – looks at the role of 'flagship' developments in the urban regeneration process through a number of case studies.
Taylor, T. (1991) *Urban Regeneration – Planning and Financial Incentives,* CPD Pack from College of Estate Management – looks at all the government initiatives, planning and financial incentives available to developers.

In addition look out for government publications or reports commissioned by the Department of the Environment or other government departments, e.g. The Audit Commission for Local Authorities in England and Wales (1989) *Urban Regeneration and Economic Development – The Local Government Dimension.* Reports produced annually by the various agencies involved in urban regeneration are also useful references.

CHAPTER 3: DEVELOPMENT APPRAISAL AND RISK

Useful textbooks covering development appraisal include:

Baum, A. and Mackmin, D. (1989) *The Income Approach to Property Valuation*, 3rd edn. Routledge and Kegan Paul, London – an established text introducing the student to the mathematics of property valuation.

The residual method of development appraisal and cash-flow techniques are critically analysed.

Darlow, C. (1988) *Valuation and Development Appraisal*, Estates Gazette, London – an established student textbook which covers development appraisal in the context of the entire development process, particularly from the financial viewpoint.

Scarett, D. (1991) *Property Valuation: The 5 Methods*, E & FN Spon, London – an introductory text which includes coverage of the residual method and cashflow techniques.

Byrne, P. *Risk and Uncertainty and Decision-Making in Property Development*, 2nd edn, E & FN Spon, London (in press) – provides an explanation of the various management and risk analysis techniques available to improve decisions made in relation to development appraisals. Provides much greater detail on sensitivity analysis and the Monte Carlo method.

Royal Institution of Chartered Surveyors (1994) *Report of the President's Working Party on Commercial Property Valuations*, RICS, London – includes the recommendations of a RICS working party looking at ways to improve the techniques of commercial property valuations in the future.

CHAPTER 4: FINANCE

Brett, M. (1991) *How to Read the Financial Pages*, 3rd edn, Hutchinson Business, London – useful beginners' reference for financial jargon and the workings of the financial and investment markets.

For further reading on the financial institutions and how they make their property investment decisions see:

Darlow, C. (1988) *Valuation and Development Appraisal*, 2nd edn, Estates Gazette, London.

Darlow, C. (1983) *Valuation and Investment Appraisal*, Estates Gazette, London.

McIntosh, A. and Sykes, S. (1985) *A Guide to Institutional Property Investment*, Macmillan, London.

Plender, J. (1982) *That is the Way the Money Goes: The Financial Institutions and Your Savings*, André Deutsch, London.

For more information on the funding techniques developed during the 1980s see:

Gooby, A.R. (1992) *Bricks and Mortals*, Century Business, London – provides, through interviews with the leading developers of the 1980s, examples of some of the innovative funding techniques used and how development schemes were funded in the 1980s

Barter, S. *et al.* (1988) *Real Estate Finance*, Butterworths, London – provides a detailed examination of how property is financed, including bank lending.

Brett, M. (1990) *Property and Money*, Estates Gazette, London – a useful guide for students on the principles of property investment and the funding of both property investments and developments.

Rodney, W. and Rydin, Y. (1989) Trends towards unitisation and securitisation in property markets, in *Land and Property Development: New Directions* (ed. R. Grover), E & FN Spon, London, pp. 81–94.

Rodney, W. (1992) *Financial Awareness*, Owlion Audio Programme (RICS and College of Estate Management) – looks at the various sources for finance for property development.

Regular reports providing up to date information on how property is being financed are provided by several sources:

DTZ Debenham Thorpe Ltd, *Money into Property*, annual.
Bank of England, *Analysis of Bank Lending*, quarterly.
Investment Property Databank, *The IPD Annual Review*, annual.
Central Statistical Office (CSO), *Financial Statistics*, quarterly.
Chesterton Financial Ltd, *Property Lending Survey*, annual.
Savills, *Property Funding Indicator*.

CHAPTER 5: PLANNING

For readers wishing to study the law and statutory procedures relating to planning and the environment see:

Moore, V. (1993) *A Practical Approach to Planning Law*, 3rd edn, Blackstone Press, London.
Ball, S. and Bell, S. (1991) *Environmental Law*, Blackstone Press, London.
Telling, A.E. and Duxbury, R.M.C. (1993) *Planning Law and Procedure*, 9th edn, Butterworths, London.

Two introductory texts for students which include an examination of the evolution and principles of planning:

Greed, C. (1993) *Introducing Town Planning*, Longman, London.
Glasson, J. (1992) *Introduction to Regional Planning*, 2nd edn, UCL Press, London.

Some useful CPD Study Packs from the College of Estate Management include:

Bather, N. (1992) *Planning Appeals* – looks at the planning appeal process.
Tuck, D. (1993) *Planning and Compensation Act 1991* – covers the changes introduced by this Act.
Comerford, J. and Stubbs, M. (1992) *Planning Gain* – looks at the way developers and local authorities negotiate agreements.

Useful sources of articles on current issues affecting development and planning include:

Planning Week
Journal of Planning and Environmental Law
The Planner
Estates Gazette

CHAPTER 6: CONSTRUCTION

Masterman, J.W.E. (1992) *An Introduction to Building Procurement Systems,* E & FN Spon, London – provides a very detailed introduction to students at an advanced level on the various procurement methods available, summarizing their advantages and disadvantages.
Chapell, D. (1991) *Which Form of Building Contract*, Longman, London – examines the various forms of building contract and when they are applicable.
Turner, A. (1990) *Building Procurement*, Macmillan, London – sets out the basics of the building process and explains the various building contracts.
Waterhouse, R. (1992) *A Guide to Project Management*, CPD Study Pack, College of Estate Management – provides a basic understanding of project management.

CHAPTER 7: MARKET RESEARCH

The following is a brief list of sources of information for property research, together with suggested further reading. The large practices produce far more information than is listed below and will normally supply a list of publications on request.

The Economy, Population and Demographics

The National Accounts (Blue Book), annual, CSO.
Economic Trends, monthly, CSO.
Financial Statistics, monthly, CSO.
Census of Employment (1981, 1987, 1989, 1991), Employment Department. Access to this, and other, employment data can be obtained via the NOMIS on-line service at the University of Durham.
The Census of Population, OPCS.
Social Trends, annual, CSO.
Family Expenditure Survey, annual, CSO.
Regional Trends, annual, CSO.
Transport Statistics, annual, Department of Transport.

Property development

Official statistics on this topic are sparse. The main sources are:

Housing & Construction Statistics, quarterly, Department of the Environment, including New Construction Orders.
Industrial and Commercial Floorspace Statistics, annual from 1974 to 1985, Department of the Environment.

In addition, local authorities can sometimes provide development statistics for their areas.

Market reports, rents yields, etc.

Property Market Report, twice yearly, Inland Revenue.
Industrial Floorspace, 6-monthly, King Sturge.
PRIME, 6-monthly, Healey & Baker.
50 Centres, 6-monthly, Jones Lang Wootton.
Property Investors' Digest, Investment Property Databank (IPD).
IPD Monthly and Quarterly Indices.
Average Yields, quarterly, Hillier Parker.
Property Index, Jones Lang Wootton.

The following firms also produce regular/occasional research reports or have an in-house research department: DTZ Debenham Thorpe, Chesterton, Gerald Eve, Grimley JR Eve, Knight Frank Rutley, Lambert Smith

Hampton, Richard Saunders & Partners, Savills. In Scotland, Ryden produce authoratative market reports. This list is expanding all the time; our apologies to any firms we have omitted.

Property databases

FOCUS, run by Property Intelligence.
APR-Glenigan Property Research.

Building cycles

Barras, R. (1994) Property and the economic cycle; building cycles revisited. *Journal of Property Research*, 11, 183–197 – as well as presenting a straightforward explanation of the relationship between the economy and building cycles, the article also references earlier relevant literature.
University of Aberdeen and Investment Property Databank (1994) *Understanding the Property Cycle*, Royal Institution of Chartered Surveyors, London – this report provides a comprehensive analysis of the facors driving building cycles and is also a source of further references.

Forecasts

Forecasts for the UK/national level include:

Forecasts for the UK Economy. A Comparison of Independent Forecasts, HM Treasury.
A Framework Forecast for the UK, Henley Centre.

Regular regional economic forecasts (but available only on subscription) are produced by:

Business Strategies Ltd, *Regional Planning Service*.
Cambridge Econometrics, *Regional Economic Prospects*.

Portfolio analysis

Brown, G. (1991) *Property Investment and the Capital Markets*, E & FN Spon, London.

References can also be found in the University of Aberdeen/IPD report on property cycles.

General

In addition to the references above, developments in all aspects of property research – applied and theoretical – are reported in the *Journal of Property Research*.

CHAPTER 8: PROMOTION AND SELLING

There is little literature on this subject but two books which specifically relate to the property industry include:

Cleaveley, E.S. (1984) *The Marketing of Industrial and Commercial Property*, Estates Gazette, London – a student textbook which includes an examination of the principles of marketing followed by advice on ways of promoting property.

Bevan, O.A. (1991) *Marketing and Property People*, Macmillan, London – provides a useful introduction to the principles of marketing for students on property related courses.

For an established textbook for students on the law relating to landlord and tenant see:

Smith, P.F. (1993) *The Law of Landlord and Tenant*, 4th edn, Butterworths, London.

POSTSCRIPT

Many interesting challenges face the property development industry as we approach the next millennium. As predicted in the last edition the industry has emerged from the boom and slump of the late 1980s and early 1990s into a property market characterized by an excess of supply and low rental growth, within the context of an economy characterized by low inflation and rapidly changing employment patterns.

There has been much discussion about developers learning from the mistakes of the past if the excesses and failures of successive booms and slumps are to be avoided in the future. However, the property development process is an imperfect one, providing a large and indivisible product over a long period of time and failing to match the shorter business cycle to which it responds. Although the wider use and availability of research has improved decision making, it cannot overcome this fundamental problem. The experience of the 1980s boom and bust has focused the attention of all the players concerned on the many weaknesses that still exist within the property development process. There is room for improvement in all stages of the process, including evaluation techniques, building contractual arrangements, the reduction of building costs and the effective use of market research.

If developers are to respond to the economic climate of the 1990s they must address the needs of occupiers in both the design and specification of buildings and the structure of leases. Fundamental changes are taking place in the way in which businesses are run with changing working patterns and rapid advances in information technology which will have an impact on property requirements in the future. These will evolve over time and developers must be able to respond quickly and effectively to the needs of occupiers. Many see a simple straightforward approach as the answer with the provision of small flexible accommodation which is efficient and inexpensive to manage, and above all, available on shorter flexible lease terms.

Developers not only face pressures from occupiers, but also from financiers and the government. The financial institutions view property in competition with all other asset classes in respect of current and predicted returns. The issues of the illiquidity of property and performance measurement techniques must be addressed if property is to effectively compete for a share of the investment allocations by the institutions. Banks remain cautious in their lending policies while recent experiences of bad debt remain fresh in their memories. New and innovative funding techniques might provide the answer.

The government is using the planning system to bring environmental considerations to the forefront of planning policy. There is much debate about the issues of contaminated land, out of town development versus the decline in many of our town centres, and the provision of roads and public transport. All these issues are inextricably linked, and one common thread is the question of who pays? There is only so much in the property development 'pot' and it must be remembered that a residual land value must exist if viable development is to proceed.

Rosalyn Topping
July 1995

INDEX

Page numbers appearing in *italic* refer to case studies.

ownership, rating and new development. Its Town FOCUS service also provides selected socio-economic and demographic indicators.

Property market information is currently delivered in printed reports or via telephone modem links which produce computer-based text or spreadsheet output. The next exciting step on the information front will be the introduction of Geographical Information Systems (GIS) into the storage, use and presentation of property data. Progress in this area is in its infancy and we look forward to reporting new developments in the next edition of this book.

(b) Supporting information

As property market analysis has evolved and become more sophisticated, it has drawn more and more on what the *Mallinson Report* calls 'supporting evidence'. As we shall see when we look at different types of analysis, however, this information is far from just 'supporting' and is often central to unravelling why the property market in general, or in a specific place, behaves in a particular manner.

For instance, if demand is analysed solely on the basis of deals completed on individual properties part of the picture may be missed: the number and type of completed deals, for example, is often restricted by the availability of suitable property. Further light can be shed on the underlying level and character of demand by looking at an area's demographic characteristics (especially for retail developments) or the structure and performance of its local economy, amongst other things. Official statistics of this type are available from the decennial Population Census and tri-annual Census of Employment; also from consultancies which provide additional analysis of Census data (e.g. CACI and URPI), local authorities and Training and Enterprise councils (TECs).

As well as indicators of local economic well-being, the analyst may also be required to advise on the overall state of the economy or particular occupier types, e.g. manufacturers or retailers. Many such macro-economic indicators are employed in property market analysis including, for example, statistics on Gross Domestic Product (GDP), retail sales and manufacturing output. These data are released by the CSO in a number of official publications, most importantly the *National Accounts* (or Blue Book), *Economic Trends* and *Financial Statistics*. Information on employment and unemployment is produced by the Employment Department. National employment statistics and selected data for regions, counties and local authorities is published in *Employment Gazette*; further geographical or industry detail is available (at a cost) from NOMIS at the University of Durham, where various Employment Department databases are held.

Other government statistics sometimes useful to property researchers can be found in, for example, *Regional Studies, Social Trends, The Family Expenditure Survey, The Labour Force Survey, Transport Statistics* and *Housing & Construction Statistics*.

The above survey of data sources (property and supporting) is by no means exhaustive and the range of useful publications is increasing all the time. Nonetheless, substantial gaps persist. All too often, researchers and analysts are spending their (and client's) time piecing together information which might much more efficiently be provided via official sources. A case in point is the almost total lack of information on commercial floorspace (although we understand the Department of the Environment may soon address this issue). Moreover, the fact that property information has developed in a piecemeal fashion, led by competing private-sector interests, means that it is often inconsistent between sources (though usually consistent within any given source) and has focused on areas which have attracted the most investment – central London and retail. This latter aspect is not altogether bad so long as the market remains focused in the same direction: the collapse of the central London office market at the end of the 1980s, however, has shifted investor and developer interests elsewhere and into locations and property types on which existing information is poor. As the next development cycle gathers pace, some new and grand mistakes may well be made in these poorly researched and understood markets.

7.4.2   Strategic and site-specific analysis

For some purposes it is often sufficient for the surveyor or researcher to supply his/her client simply with information. As information has become more widely available, however, clients have, quite rightly, come to expect interpretation of the information in the light of specific questions about their company or individual development plans. Here, we outline briefly two different types of analysis which are commonly required:

a. Strategic analysis
b. Site-specific research.

(a) Strategic analysis

Strategic analysis is conducted either by a company's own research team or commissioned from external consultants. It typically examines questions which are long term in nature and rarely relates to individual development projects.